Laboratory Experiments for General Chemistry

Fifth Edition

Toby F. Block
Georgia Institute of Technology

George M. McElvy
Georgia Institute of Technology

Australia • Canada • Mexico • Singapore • Spain • United Kingdom • United States

Publisher: David Harris
Acquisitions Editor: Lisa Shaver
Assistant Editor: Ellen Bitter
Marketing Manager: Amee Mosley
Marketing Assistant: Michele Collela
Project Manager, Editorial Production: Belinda Krohmer
Permissions Editor: Kiely Sisk
Print Buyer: Lisa Claudeanos
Printer: P. A. Hutchison

Printed in the United States of America
1 2 3 4 5 6 7 08 07 06 05 04

Printer: P.A. Hutchison

0-534-42448-1

For more information about our products, contact us at:
Thomson Learning Academic Resource Center
1-800-423-0563

For permission to use material from this text or product, submit a request online at **http://www.thomsonrights.com.**
Any additional questions about permissions can be submitted by email to **thomsonrights@thomson.com.**

Thomson Higher Education
10 Davis Drive
Belmont, CA 94002-3098
USA

Asia (including India)
Thomson Learning
5 Shenton Way
#01-01 UIC Building
Singapore 068808

Australia/New Zealand
Thomson Learning Australia
102 Dodds Street
Southbank, Victoria 3006
Australia

Canada
Thomson Nelson
1120 Birchmount Road
Toronto, Ontario M1K 5G4
Canada

UK/Europe/Middle East/Africa
Thomson Learning
High Holborn House
50–51 Bedford Row
London WC1R 4LR
United Kingdom

Latin America
Thomson Learning
Seneca, 53
Colonia Polanco
11560 Mexico
D.F. Mexico

Spain (including Portugal)
Thomson Paraninfo
Calle Magallanes, 25
28015 Madrid, Spain

Preface

As *Laboratory Experiments for General Chemistry* enters its fifth edition, we, the authors, are proud to say that the manual has been used successfully in high schools, two- and four-year colleges, and technical institutes both across the United States and abroad over a span of more than a decade. While we are gratified to have heard many positive comments from the adopters of the book, we know that even a good book must change as trends in education evolve and as the needs of students and teachers are continually modified. Therefore, while we have maintained many features of past editions, we have also introduced many new features in the fifth edition.

Pre-revision reviews revealed that a number of the experiments in the fourth edition were no longer being performed at many of the schools that had adopted *Laboratory Experiments for General Chemistry*. Therefore, we have eliminated eighteen experiments, including three experiments that dealt with classical qualitative analysis, three experiments (Atomic Energy Levels and Atomic Spectra, Constructing an Alien Periodic Table, and Isomerism in Organic Chemistry) that did not involve any real laboratory work, and several experiments (Acid/Base Titration of Ascorbic Acid, The Vapor Pressure of an Azeotropic Mixture of Isopropyl Alcohol and Water, Determination of the Solubility Product of Copper(II) Iodate, Kinetics, Chemical Equilibrium) that were felt to be somewhat redundant in that the principles and skills involved in their performance were covered adequately by related experiments that have been retained in the fifth edition. (Of course, in today's diverse world, it may very well be the case that an experiment that was not being used by many departments was very well–suited to the needs of some institutions. Anyone who wishes to continue using experiments that no longer appear in the manual should contact the custom publishing department at Thomson Learning.)

Experiments that have been retained include Getting Started in the Laboratory; On the Nature of Pennies; Properties of Oxides, Hydroxides, and Oxo–acids; The Burning of a Candle; Temperature Change and Equilibrium; and Electrochemical Cells. Some of the retained experiments have been modified — additional calculational support has been provided in Volumetric Analysis: Acid/Base Titration Using Indicators and Determination of the Composition of Cobalt Oxalate Hydrate. Quantities of reagents have been altered in Determination of the Composition of Cobalt Oxalate Hydrate and Thermodynamic Prediction of Precipitation Reactions to facilitate the completion of the experiment within the three-hour laboratory period and to make the outcomes of the reactions more easily observable. Some experiments, such as Kinetic Study of the Reaction between Iron(III) and Iodide Ions, have been extensively rewritten to give students insight into their complex experimental set-ups. We have also greatly expanded our treatment of mathematical manipulations in Appendix A and included extensive instruction on the use of computer-based graphing procedures.

Several experiments (Getting Started in the Laboratory; Volumetric Analysis: Acid/Base Titration Using Indicators; Vapor Pressure of Water; Freezing-Point Depression; Determination of the Dissociation Constant of an Acid/Base Indicator; and Electrochemical Cells) are presented in two formats — the "traditional" method and a "probe-based" method. Others (Freezing-Point Depression and Iodometric Analyses) offer a choice of systems for study.

Of course, we have retained those features of the previous editions that we and other adopters of the manual felt to be the strong points of the book. Each experiment begins with Safety precautions and suggestions for rendering First Aid and ends with directions for the safe disposal of reagents. We have retained the "Objective" and "Procedure in a Nutshell" features, introduced in the fourth edition; reviewers have reported that these sections have increased student's ability to focus on both the procedural aspects of the experiments and the goals of the exercises. Many of the experiments are best characterized as guided inquiry exercises in which students are given information and guidance, but are also allowed the freedom to explore and are, indeed, required to make decisions on the quality of their results and the possible need for further experimentation. Several experiments can be conducted as collaborative efforts; these are designated by a special icon in the Table of Contents and instructions for the facilitation of teamwork are given in either the manual itself or in the Instructor's Guide. The experiments are done in microscale where practical, often incorporating household materials and keeping the use of hazardous materials to a minimum.

As with our previous editions, we have striven to produce a book that could be used by students who differ in scientific attitude, interest in chemistry, and scholastic background. We have attempted to design experiments that require students to work carefully to obtain good data and that also require students to put some thought into the calculations needed to convert good data into good results. We have introduced more collaborative work partly in response to the fact that most students at Georgia Tech now take only one semester of chemistry. However, we also appreciate that group work can be valuable even in settings in which class size is smaller and more time is allotted to the course's laboratory work. It teaches the importance of cooperation, the need to report data accurately and in full detail, and the absolute necessity of both learning to draw conclusions from the work of others and to determine the degree of reliance one can have on such conclusions.

Although we have deliberately chosen not to replace hands-on, wet chemistry with computer simulations, we recognize that computers play a significant role in modern education. Pre-Laboratory Videos and Quizzes have been successfully adapted from the lab manual and placed on-line for access via the World Wide Web and, in our case, WebCT™, by one of the authors (GMM). In-house research has shown that students who use these tools on the web come to the laboratory period better prepared, place fewer demands on the time of the laboratory instructor, spend more time on-task, and, in general, maintain a safer laboratory. In addition, some of the experiments' Summary Report Sheets have been encoded into

HTML, allowing submission of laboratory data via e-mail to class members for inclusion in collaborative exercises. These too are presented via WebCT™. For more information, please contact George McKelvy.

We hope that users of previous editions of the manual will find the latest revision well-suited to their needs. We welcome their comments as well as those of new adopters of the manual.

Acknowledgments

We are indebted to Myung-Hoon Kim (Georgia Perimeter College-Dunwoody), Mary Leigh Lyon (Holmes Community College), Paulos Johannes (Georgia Perimeter College-Lawrenceville), and Brentley S. Olive (University of North Alabama) for their insightful comments during the pre-revision review process. We would like to thank Mr. Hongjin Jiang for his work in testing and perfecting the CBL methods described in Experiments 1P, 15P, 16P, and 19P and Mr. Tatsuya Maehigashi for preparing the new graphing guidelines for Appendix A. We would also like to express our appreciation to our spouses, Jerrold Greenberg and Julie McKelvy, for their support during the preparation of this manuscript and to Annie Mac and Ellen Bitter for their efforts in making the fifth edition of *Laboratory Experiments for General Chemistry* a reality.

Toby F. Block
George M. McKelvy
Atlanta, GA
December, 2004

Table of Contents

[*] Experiments making use of the Vernier CBL-2 and assorted probes are included as supplements or addendums to selected experiments.

* Experiments making use of the Vernier CBL-2 and assorted probes are included as supplements or addendums to selected experiments.

Denotes availability as Group Experiment

* Experiments making use of the Vernier CBL-2 and assorted probes are included as supplements or addendums to selected experiments.

Introduction: Laboratory Conduct and Procedures

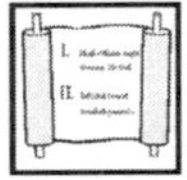

General Rules and Suggestions

You are in the laboratory primarily to learn by doing experiments. You will benefit most from the experience if you have prepared in advance by reading the assigned experiment and completing its pre-lab exercises. You are likely to waste precious laboratory time if you attempt to perform experiments without appropriate preparation. Always be sure to review (before lab) the sequence of operations to be performed in the assigned experiment and note those procedures that are time-consuming or hazardous. Then plan your work so that you can complete the experiment in a timely, safe manner.

Keep in mind that you share the laboratory and its equipment with other students. Therefore, you should work safely and considerately. Loud noises, practical jokes, and similar offensive behavior have no place in the chemistry laboratory.

Do not attempt to repair or adjust malfunctioning instruments. Report problems to your instructor, who will see to it that instruments are repaired or adjusted.

Before you leave the laboratory, clean your work area. This includes your place at the bench, the reagent shelf, the sink, and other areas shared by the class. Dispose of waste material in an approved manner, following the instructions provided with each experiment.

Safety

In the course of your laboratory work, you will be exposed to new instruments, new procedures, new chemicals, and, possibly, new hazards to health and safety. Fortunately, most laboratory accidents are avoidable. You can minimize the possibility of injury by observing the general safety rules given below. In addition to these rules, you should pay close attention to the discussion of possible hazards given in each experiment, and you should follow all procedures carefully. Study the First Aid section and be prepared to administer prompt and appropriate first aid should you or another student be injured.

1. Wear appropriate eye and face protection (safety goggles and/or face shield) whenever you are in the laboratory. This rule applies even when you are not conducting experiments; you could be the victim of another person's carelessness.

2. At the very beginning of the term, learn the location and proper use of the fire extinguishers, safety shower, eye-wash fountains, and other emergency equipment that is found in your lab. This will permit you to use such equipment without delay in an emergency. You should also locate the exits from the lab and plan an escape route to be used in case of fire.

3. Do not perform unauthorized experiments. Do not work in the laboratory unless an instructor is present.

4. Regard all chemicals as hazardous; do not touch, taste, or intentionally inhale chemicals.

5. Do not eat, drink, or smoke in the laboratory. This will prevent the accidental ingestion of chemicals.

6. Never mouth pipet; always use a pipet bulb.

7. Do not use burners or open flames where acetone or other flammable solvents are being used.

8. Be sure beakers or flasks used for boiling liquids are securely supported by a tripod or iron ring. Use boiling chips to avoid bumping. When boiling small amounts of liquid in a test tube, always aim the open end of the tube away from yourself and other persons.

9. Do not use glass tubing unless the cut ends have been fire polished to round their sharp edges. When inserting glass tubing or thermometers into rubber stoppers, lubricate the glass first with glycerin or water. Grasp the glass near the end to be inserted to minimize the chance of breakage. Use a towel between the glass and your hand to protect yourself from cuts—in the event that the glass breaks. Chipped or broken glassware of any kind should be replaced with new equipment.

10. Never add water to concentrated acids or bases because the heat of dilution may cause the corrosive solution to splatter. Dilute acids or bases by slowly adding small amounts of concentrated reagent to a large amount of water, with constant stirring.

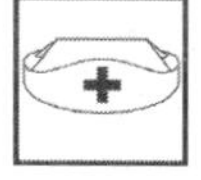

First Aid

It is your responsibility to know the location of the emergency equipment. You should also know how to use such equipment. General first-aid recommendations are given in this section. Specific information concerning serious injury appears at the start of each experiment.

If you splash acids, bases, or other chemicals on yourself or in your eyes, wash the affected area immediately with large quantities of water, first giving attention to your eyes. Continue washing until all corrosive materials have been removed. The eyelid should be pulled up and away from the eye, by another person if necessary, while the eye is thoroughly irrigated with plain water. This procedure is most easily accomplished at an eyewash fountain. After chemicals have been rinsed away from the eye, seek medical attention immediately, to guard against permanent vision impairment.

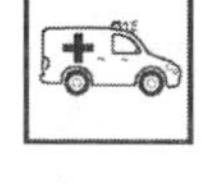

Should your clothing catch on fire, remain as calm as possible. Walk (do not run) to the safety shower and pull the ring to douse yourself with water. Alternatively, you may drop to the floor and roll to extinguish the flames.

Minor burns are best treated by soaking the affected area in cool water. Minor cuts can be covered with adhesive bandages. Report such injuries to your instructor, who will assist you in treating them. (Because these injuries occur so frequently, both in and out of the laboratory, we will not discuss their treatment in the First Aid sections of the experiments but will instead focus on more serious injuries that might occur.)

Accidents are usually preventable. Be alert in the lab. Understand the experimental procedures and be aware of any hazards. Ask your instructor to explain operations you do not understand. Do not proceed until you know what you need to do.

Waste Disposal

As recently as 25 years ago, academic laboratories disposed of their chemical wastes in the most convenient manner: volatile solvents were allowed to evaporate; solids were dumped into the general trash; liquids and solutions were flushed down the sink. These practices are unacceptable today. There are now rules for disposing of hazardous waste. Unfortunately, these rules are often difficult and expensive to follow. Thus, it is desirable to minimize the amount of hazardous wastes generated in academic laboratories. We have incorporated the following procedures into these experiments to accomplish this.

1. We have reduced the scale of experiments.

2. We have replaced hazardous chemicals with less hazardous (or nonhazardous) materials.

3. We have given instructions for converting wastes to nonhazardous form, by means of chemical reaction, prior to disposal.

4. We have given instructions for reducing the volumes to be disposed of, by concentrating solutions or by separating hazardous chemicals as precipitates.

5. We have given instructions for recovering chemicals for reuse, whenever practical.

Special instructions for the disposal of wastes are provided with the experiments. In most cases, the wastes generated in these experiments are nonhazardous and, unless local ordinances specify otherwise, can be rinsed down the sink according to the guidelines given below.

1. Solutions acceptable for sink disposal should be diluted one hundredfold with water before being slowly poured down the drain. This should be followed by flushing with fresh water. Alternatively, the solution may be poured slowly down the drain along with a stream of water sufficient to accomplish the desired dilution.

2. Unless otherwise specified, soluble salts may be disposed of as described in Step 1.

3. Small amounts of water-soluble organic solvents (such as methanol or acetone) may be diluted and flushed down the drain. Large amounts of these solvents or any amount of volatile, highly flammable solvents (such as ether) should not be disposed of in this way.

4. Solutions of acids and bases should have their pH adjusted to the 2–11 range before being flushed down the sink. Small amounts (10 mL or so) of dilute acid or base may be simply diluted and flushed.

When in doubt, check with your instructor before pouring any chemical down the drain.

The Notebook and the Report

Rules for Keeping the Notebook or Making Reports

Your instructor will determine if you are to maintain a laboratory notebook or if you will be submitting individual reports on your experiments. If a notebook is used, it will serve as an original, permanent record of your work. Thus, a permanently bound notebook will be required. Whether you maintain a notebook or not, you should follow the general guidelines given below for report preparation. You should also follow any additional directions given by your instructor.

If You Keep a Notebook

1. The first two or three pages of the notebook should be left blank because they will be used to prepare a Table of Contents. The Table of Contents should be updated as each report is completed. If the notebook pages are not numbered, you should number them consecutively in ink.

2. All experimental observations should be recorded in ink directly in the notebook at the time of observation. Do not record figures on loose scraps of paper to be later copied neatly into the notebook. You may very well lose such scraps of paper and, even if you don't, copying destroys the value of the notebook as the original record of your laboratory work.

3. Do not obliterate or erase any data recorded in the notebook. If you wish to correct an error, draw a single line through the incorrect number or statement, and then write the correct entry beside it. If the reason for the change is not obvious, write a brief explanation in parentheses near the corrected data.

4. Do not remove pages from the notebook.

5. Try to be neat. Study the experiment and think it through in advance. Then decide on the clearest and most efficient way of recording your observations and make any necessary advance preparations, such as designing a table to be filled in at the laboratory. Some students strive for neatness by writing their formal reports, using only the right (or left) side of the notebook pages, leaving the other side for sketches and calculations.

If You Do Not Keep a Notebook

All experimental observations should be recorded directly on the Summary Report Sheets that accompany each experiment. Do not record figures on loose scraps of paper to be later copied neatly onto the Report Sheet. You may very well lose such scraps of paper and, even if you don't, copying destroys the value of the Summary Report Sheet as the original record of your laboratory work.

Do not obliterate or erase any data recorded on your Summary Report Sheet. If you wish to correct an error, draw a single line through the incorrect number or statement, and then write the correct entry beside it. If the reason for the change is not obvious, write a brief explanation in parentheses near the corrected data.

Whether or Not You Keep a Notebook

The report for each experiment should contain each of the sections listed below, arranged in the order presented.

1. *The TITLE of the experiment.*

2. *The PURPOSE of the experiment.* Here you should list the skills to be learned or the principles to be demonstrated by proper performance of the experiment.

3. *The PROCEDURE followed.* Briefly describe the equipment used and the procedure followed. It is not necessary to list the minor details of the procedure unless a new procedure has been substituted (with your instructor's approval) for the one given in the laboratory manual.

4. *The original DATA.* Data should be recorded directly in the notebook or on the Summary Report Sheets.

5. *Tabulated RESULTS.* In this section, you should give a table, or tables, showing all significant results obtained from your experiment.

6. *DISCUSSION.* In this section, you should show sample calculations relating your data to your results. Briefly discuss any errors that were made in the course of the experiment. You should also discuss the significance of the experimental results.

7. *QUESTIONS.* Many of the experiments in this laboratory manual contain questions for you to answer. Answer these questions as part of the report.

Because you have a limited amount of time in the laboratory in which to complete each experiment, as much of the report as is practical should be written outside the laboratory. The Title, Purpose, and Procedure sections may be written before the laboratory period. However, you must record all data (in the notebook or on the Summary Report Sheets) only at the time the original observations are made in the laboratory. You may complete the Results, Discussion, and Questions sections either in the laboratory or at home, depending on the time available.

Laboratory Equipment

Common items of laboratory equipment are shown in Figures I.1 through I.3. Figure I.1 shows commonly used items of glassware. Figure I.2 shows other laboratory equipment. Study the figures until you are able to identify any unfamiliar items.

The Balance and Weighing

Despite the difference between mass and weight, we usually use the term "weighing" for the operation of determining the mass of an object using a laboratory balance. In weighing, we add or remove standard masses (often called "weights") until the gravitational attraction for the weights is equal to the gravitational attraction for the object. The mass of the object is then equal to the sum of the standard weights.

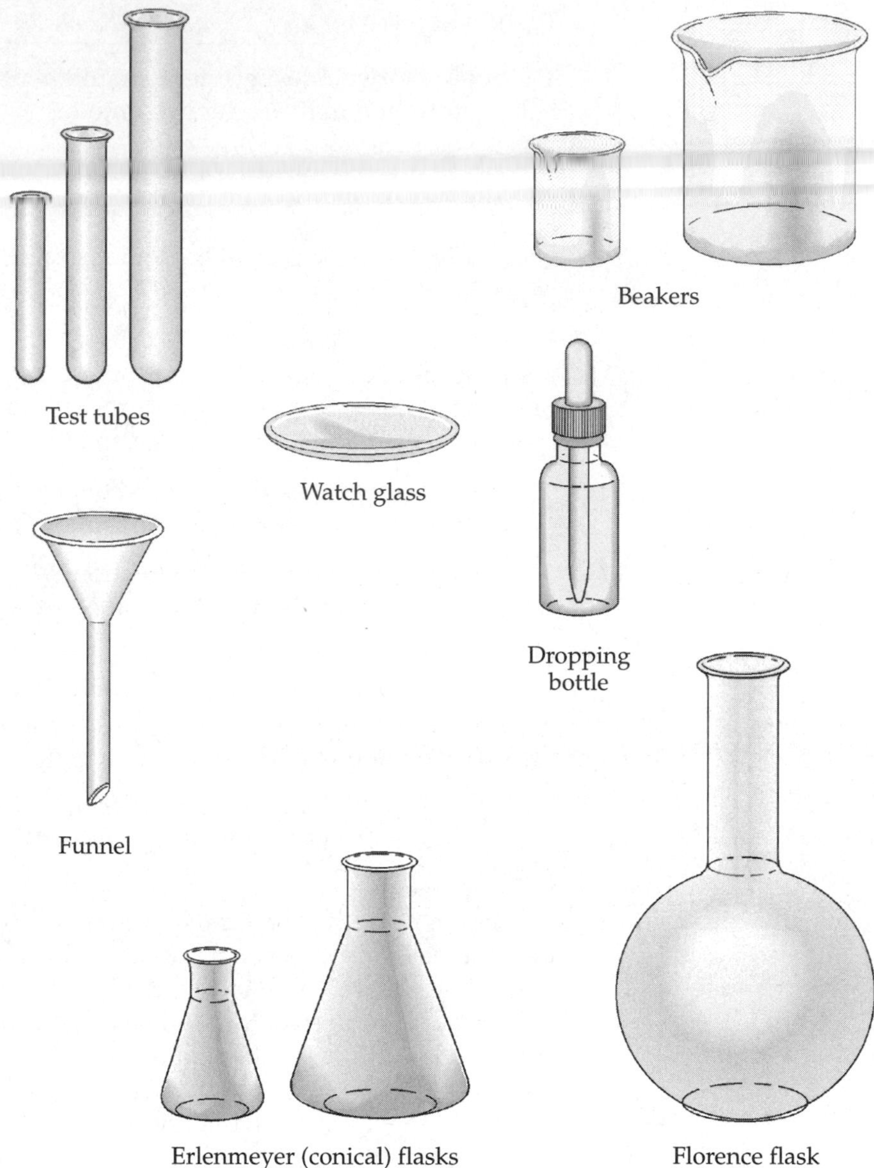

Figure I.1 General glassware

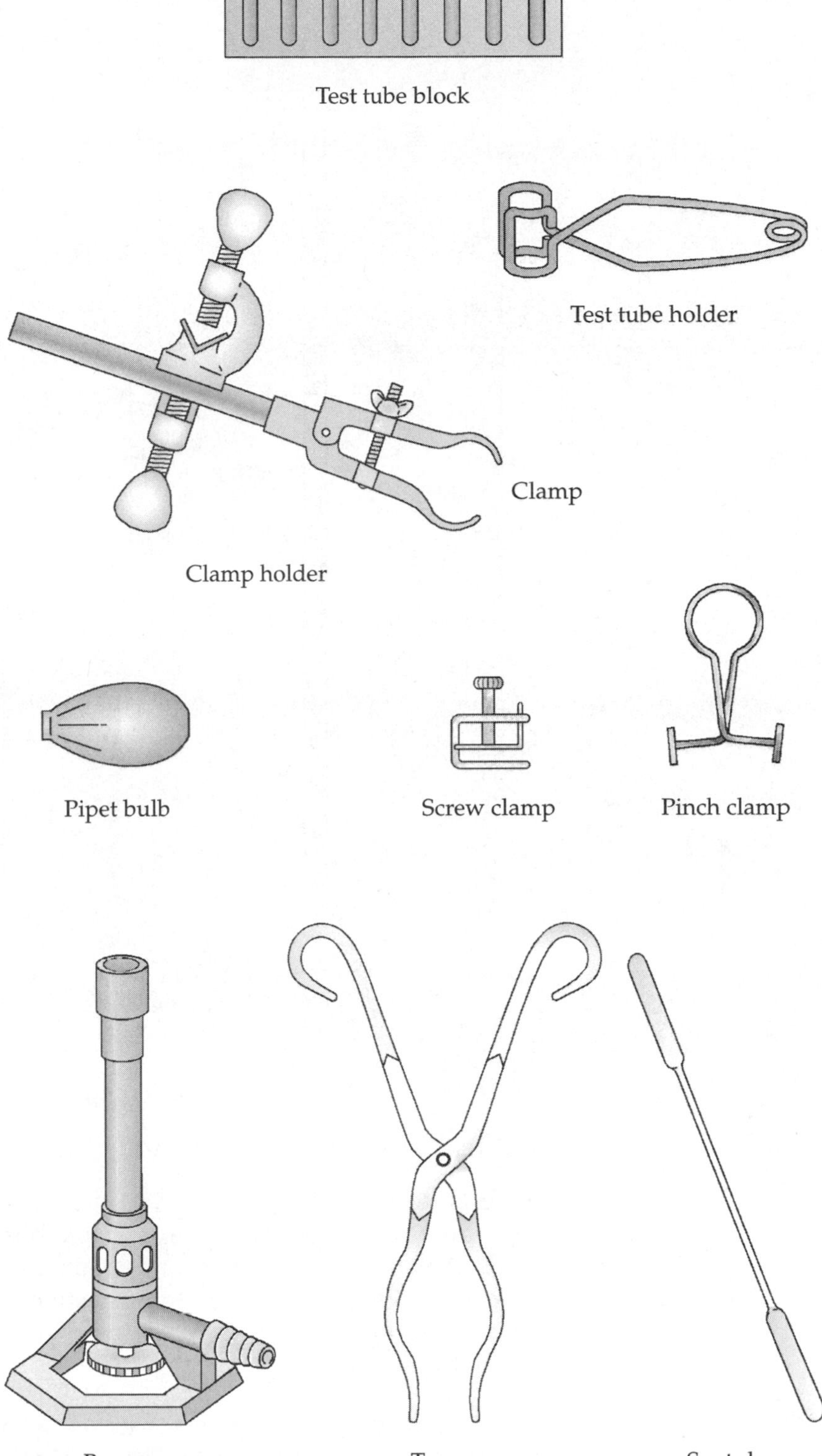

Figure I.2 Other equipment

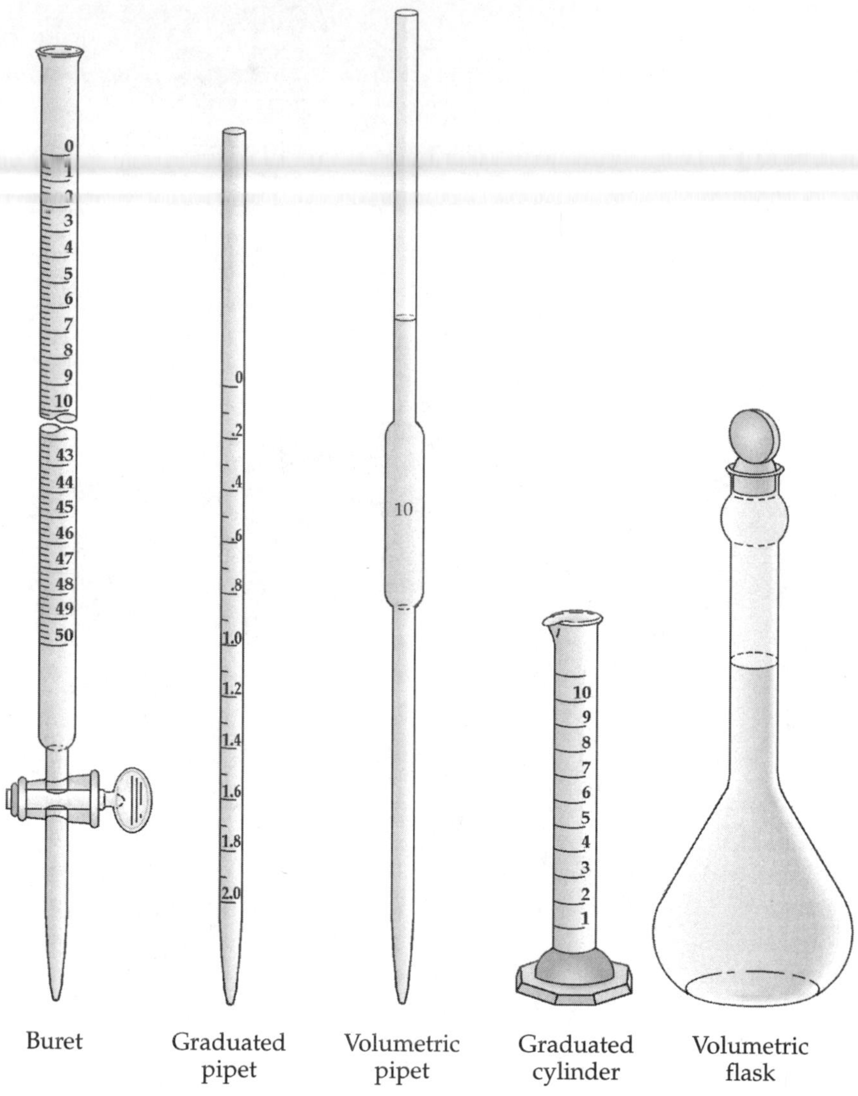

Figure I.3 Volumetric glassware

Rules for Using Laboratory Balances

Several different types of laboratory balances are used in general chemistry laboratories. These range from triple-beam balances to mechanical analytical balances to electronic analytical balances with digital readout. Your instructor will provide you with directions for using the types of balances found in your lab. However, the following general rules are appropriate for all balances.

1. Do not attempt to repair or adjust a malfunctioning balance. Inform the instructor, who will see to it that the problem is corrected.

2. Never place chemicals on the balance pans. Solids should be weighed in a tared (pre-weighed) beaker, weighing boat, or watch glass. Liquids should be weighed only in a stoppered container, such as a flask or weighing bottle.

3. If you spill chemicals in or around the balance, use a balance brush to clean up the mess before you leave the balance. It is very important to remove all traces of spilled chemicals from inside the balance case because chemicals may corrode the balance and ruin it. Your instructor may be able to help you identify the spilled chemicals and can recommend an appropriate disposal procedure.

4. If the balance being used has knobs for adding or removing weights, turn the knobs slowly and return them to the zero setting before leaving the balance.

5. Do not weigh hot or cold objects.

6. When recording masses, do not lean on the balance table; learn to support your notebook or laboratory manual on your lap.

Volumetric Glassware

The most common items of equipment designed for measuring the volumes of liquids are illustrated in Figure I.3. Look at these items carefully and note the permanent markings provided to aid you in using the item correctly. In addition to their calibration marks, volumetric glassware is often marked to show the calibration temperature and intended method of use. The symbols "TC" and "TD" marked on a piece of volumetric glassware indicate, respectively, that the apparatus has been calibrated to contain or to deliver the specified volume of liquid. For example, a volumetric flask will be designated TC, a buret or pipet will be marked TD, and a graduated cylinder may be marked TC or TD, depending on its intended application. The volumetric pipet and volumetric flask have a single calibration mark and can be used for measuring only the specified volume of liquid; the graduated cylinder, measuring pipet, and buret are marked with a scale that permits volumes from 0 mL up to the capacity of the apparatus to be measured.

Cleaning Glassware

Before using a pipet, buret, or other volumetric item, you should first fill it with distilled water and then allow the water to drain out completely under the influence of gravity. If the glass is clean, only a thin, even film of water will remain on the walls. Where there is grease or dirt on the glass, large irregular drops of water will be observed. Because the calibration cannot take adhering drops into account, it is necessary to clean the glass before accurate measurements can be made.

Glassware will generally remain clean if it is washed and rinsed immediately after use, before chemicals have had an opportunity to dry on the glass. When cleaning is required, soak the item in aqueous detergent solution, then rinse it in ordinary water and then in distilled water. You may use a soft brush with graduated cylinders or volumetric flasks, but do not use a brush with expensive burets because the wire shaft of a long, stiff brush may scratch the inner walls of the buret.

Use of the Pipet

The procedure for filling a pipet by means of a rubber bulb is illustrated in Figure I.4. Hold the pipet near its upper end with your index finger in position to be moved easily to cover the open upper end of the pipet after it has been filled with liquid and the bulb has been removed. By varying the pressure of your finger against the opening, you can either hold the liquid in the pipet indefinitely or dispense it at a controlled rate. Hold the rubber bulb firmly against the end of the pipet when drawing up the liquid by suction. Do not force the bulb over the end of the pipet, as this enlarges the hole in the bulb and tends to make it unusable.

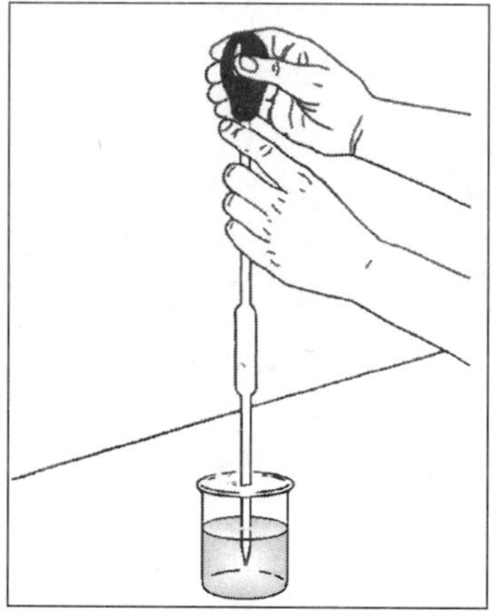

A. Use suction from a rubber bulb to draw liquid into the pipet.

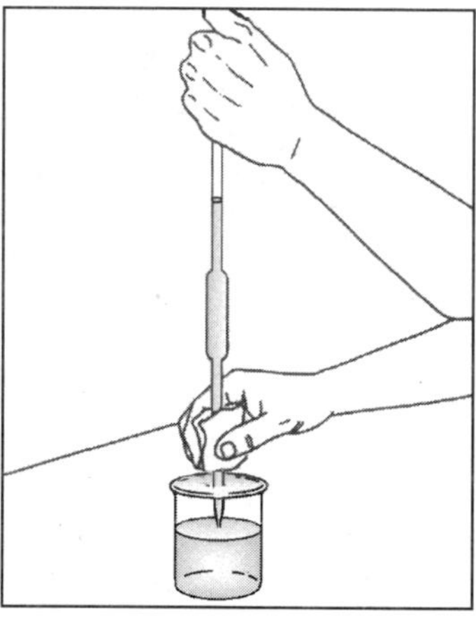

B. Dry the outside of the pipet using a tissue.

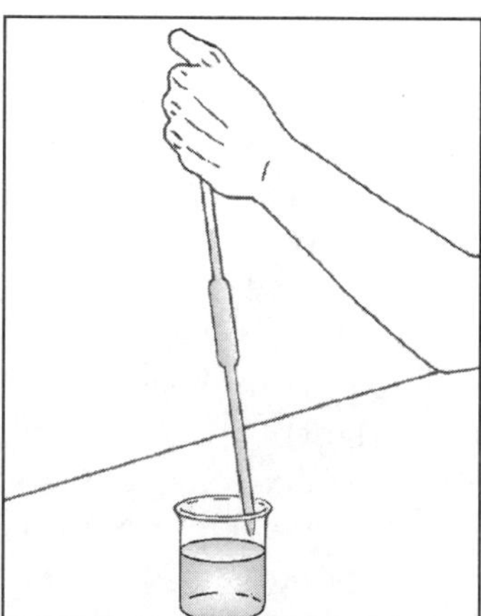

C. Carefully drain the pipet to the calibration mark.

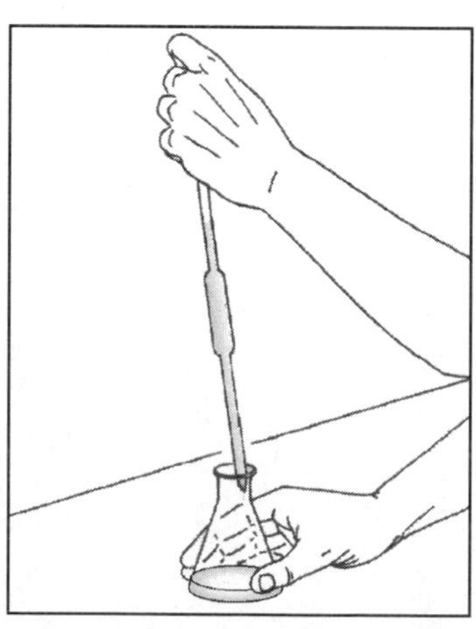

D. Transfer the pipet to the receiver and let the liquid drain out under the influence of gravity.

Figure I.4 Pipet technique

To pipet a liquid sample accurately, without diluting it or contaminating it, you should first rinse the pipet once or twice with small portions of the liquid. To do this, partially fill the pipet with the sample, then tilt and turn the pipet until the inner walls have been well rinsed with the liquid. Next drain and discard the contents of the pipet. After you have rinsed the pipet, fill it to a level approximately 1/2 inch above the calibration mark, then seal the pipet with your finger. Remove the pipet tip from the beaker or bottle containing the sample liquid; then dry off the outside of the pipet using a disposable paper tissue. Next touch the pipet tip against the side of a beaker or other waste receptacle and carefully drain out the excess liquid, stopping when the bottom of the curved liquid surface just rests on the calibration mark. Finally, move the pipet to the beaker or flask provided for collecting the sample and, once more, hold its tip against the walls while the liquid drains out under gravity. When using a measuring pipet, stop the flow when the desired calibration mark is reached. When using a volumetric pipet, let the liquid drain out naturally. A drop may remain in the tip. Do not blow this out; it has been accounted for in calibrating the pipet.

Use of the Buret

The stopcock. Most burets have horizontal Teflon stopcocks. Such burets are stored with the stopcock loosened to avoid warping of the Teflon. The first step in using such a buret is to tighten the stopcock. This is accomplished by three small items that must be found in the proper orientation: a washer next to the glass, followed by an O-ring, and, finally, a nut. It is the nut that is adjusted so that the stopcock fits snugly into the barrel of the buret. The stopcock should turn easily but should not leak. When you are finished with the buret for the day, loosen the stopcock nut before putting the buret away.

Filling the buret. Before filling the buret completely, rinse it with two 5-mL portions of the titrant solution. Then fill the buret almost to the top and turn the stopcock briefly to dispense a vigorous stream of liquid. This procedure should suffice to sweep the air out of the stopcock and dispensing tip, but it may be repeated if necessary. (Sometimes cleaning the stopcock aids in the removal of persistent bubbles. If you do remove the stopcock from the buret for cleaning, be sure that the washer, O–ring, and nut are replaced in the proper order.) Once all air bubbles have been removed from the buret, refill it and then use the stopcock to adjust the liquid level to the zero mark. Practice controlling the stopcock with your left hand (if you are right-handed) or with your right hand (if you are left-handed). Although this might feel awkward at first, this procedure leaves your more skillful hand free for swirling the titration flask.

Reading the buret. Figure I.5 illustrates the procedure for reading the liquid level in the buret. The bottom of the curved liquid surface, or meniscus, is compared with the marked scale by sighting with the eye at the same elevation as the meniscus. (You can hold a small black rectangle behind and just below the meniscus to darken it and make it more distinct.) Be sure that your eye level is not appreciably above or below the meniscus, or an error due to parallax will result. Read the volume to the smallest marked scale subdivision and obtain an additional figure by estimating the position of the meniscus between subdivisions. For example, when you use a buret marked at 0.1-mL intervals, you can estimate the liquid level to ±0.01 mL. Always be certain to record both the initial and the final buret readings because the difference between these figures is the volume of liquid dispensed.

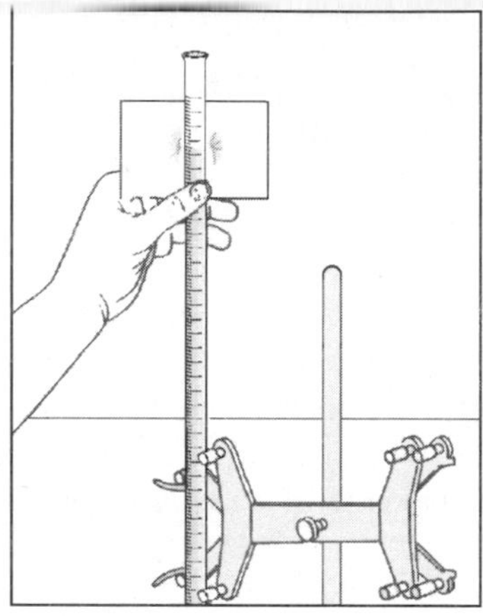

A. Read the initial liquid level using a marked card to darken the meniscus.

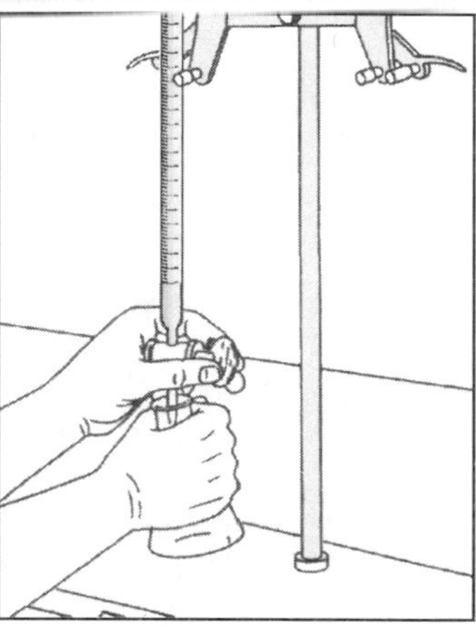

B. Control the stopcock with your left hand while swirling the flask with your right hand.

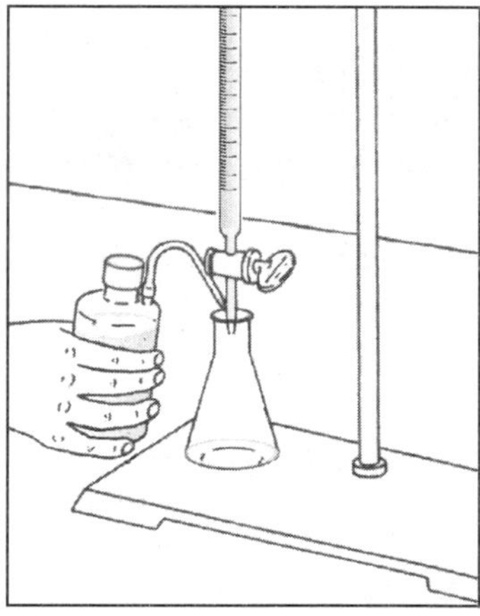

C. Rinse the walls of the flask and the buret tip when the end point is near.

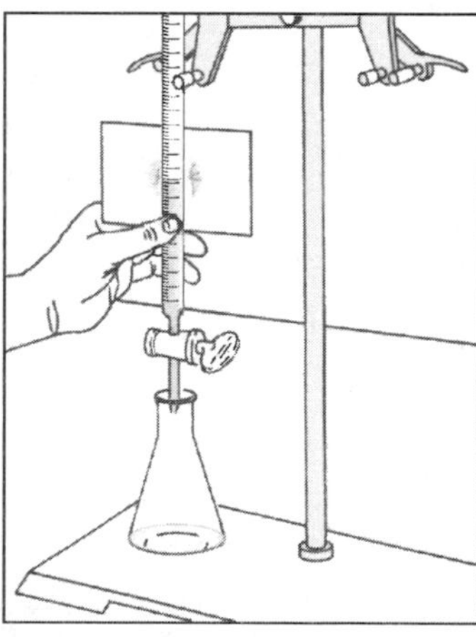

D. At the end point, read the final liquid level using the marked card to darken the meniscus.

Figure I.5 Titration technique

Reporting Data Accurately

Significant Figures

Significant figures are all the digits in a number except for any zeros written only for the purpose of locating the decimal point. The numbers 30.2, 0.00357, and 35.8 all have three significant figures. The number 35 has only two significant figures, but the number 35.0 has three because the final zero is not necessary for establishing the decimal point, and it must therefore be significant. Accepted practice is to report all accurately known digits plus one that has been estimated or is otherwise subject to uncertainty. One generally assumes an uncertainty of at least ±1 in the last digit reported. Reporting too few or too many significant figures gives a false picture of the accuracy of the measurement.

When you combine numbers, your answer should retain only one digit that is uncertain, in addition to all accurately known digits. This may require "rounding off" to the correct number of significant figures. The rules for rounding off are illustrated below.

1. *Rule:* If the first dropped figure (underlined in the example) is less than 5, drop excess figures without changing previous figures.

   Example: Round off $26.80\underline{3}7$ to four significant figures.

   Answer: 26.80

2. *Rule:* If the first dropped figure (underlined) is greater than 5, increase the previous digit by 1 unit, then drop the excess figures.

   Example: Round off $26.803\underline{7}$ to five significant figures.

   Answer: 26.804

3. *Rule:* If the dropped digit equals 5, round off the previous digit to the nearest even number.

   Example: Round off $26.82\underline{5}$ and $26.81\underline{5}$ to four significant figures.

   Answer: 26.82 in both cases.

4. *Rule:* When adding or subtracting, round off the result to the same number of decimal places as your least-precise quantity.

   Example: $15.80\underline{6} + 7.2\underline{1} + 5.\underline{1} = 28.116$ (round off to 28.1)

   The uncertain digit in each number has been underlined. Note that 28.116 incorrectly reports three uncertain figures, but only one uncertain figure is retained when we round off to 28.1 in accordance with this rule.

5. *Rule:* When multiplying or dividing numbers, report the answer to the same number of significant figures as the least precise number used.

   Example: The product of $4.2\underline{7}$ and $0.6\underline{1}$ is 2.6657. Because the least-precise number, 0.61, has two significant figures, the answer should be rounded off to 2.7.

Experimental Uncertainties

As noted above, scientific data are reported as numbers consisting of accurately known digits plus one digit that is associated with some degree of uncertainty. A mass of 6.20 g ± 0.03 g may have a true value between 6.17 g and 6.23 g. A volume reported as 10.00 mL would have a minimum true value of 9.99 mL and a maximum true value of 10.01 mL. Any result derived from experimental data, such as the density of an object with a mass of 6.20 g ± 0.03 g and a volume of 10.00 mL ± 0.01 mL, must contain at least as high a degree of uncertainty as the original data. This section will deal with ways of determining the uncertainties to be associated with results obtained from data.

In the case of our object, the reported density would be (6.20 g/10.00 mL =) 0.620 g/mL. We could calculate the uncertainty associated with this value by considering that the density will be at its minimum value when the true value of the mass is at a minimum and the true value of the volume is at a maximum. Thus, the minimum value of the density is (6.17 g/10.01 mL =) 0.616 g/mL. Likewise, the density will be at its maximum value when the true value of the mass is at a maximum and the true value of the volume is at a minimum. Thus, the maximum value of the density is (6.23 g/9.99 mL =) 0.624 g/mL. Therefore, the density of the object should be reported as 0.620 g/mL ± 0.004 g/mL.

In order to avoid doing three calculations instead of one, whenever uncertainties are to be reported along with results, you may use the following rules.

1. *Rule:* When data are to be added or subtracted, add the uncertainties.

   Example 1: Find the length of the perimeter of a square that measures 2.52 cm ± 0.02 cm on a side.

   Answer: 10.08 cm ± 0.08 cm

   Example 2: Find the mass of a sample placed in a flask, if the empty flask weighed 23.6154 g ± 0.0001 g and the flask plus sample weighed 26.2198 g ± 0.0001 g.

   Answer: 2.6044 g ± 0.0002 g

2. *Rule:* When data are to be multiplied or divided, add the relative uncertainties.

   Example 3: Find the density of an object that has a mass of 6.20 g ± 0.03 g and a volume of 10.00 mL ± 0.01 mL.

   Answer:
   $$0.620\ \text{g/mL} \pm (0.620\ \text{g/mL})\left(\frac{0.03\ \text{g}}{6.20\ \text{g}} + \frac{0.01\ \text{mL}}{10.00\ \text{mL}}\right)$$
   $$= 0.620\ \text{g/mL} \pm 0.004\ \text{g/mL}$$

   Example 4: Find the volume of a rectangular solid that measures 25 mm × 45 mm × 65 mm. Assume the uncertainty in each dimension is ± 3 mm.

Answer:

$$7.3\times10^4\ \text{mm}^3 \pm \left(7.3\times10^4\ \text{mm}^3\right)\left(\frac{3\ \text{mm}}{25\ \text{mm}}+\frac{3\ \text{mm}}{45\ \text{mm}}+\frac{3\ \text{mm}}{65\ \text{mm}}\right)$$

$$=7.3\times10^4\ \text{mm}^3 \pm 1.7\times10^4\ \text{mm}^3$$

$$=7\times10^4\ \text{mm}^3 \pm 2\times10^4\ \text{mm}^3$$

Note that, in the examples given under Rule 2, the relative uncertainty is obtained by dividing the uncertainty (e.g., ±0.01 mL) by the quantity (10.00 mL) with which it is associated. Relative uncertainties are unitless quantities. Thus, relative uncertainties for different types of data (e.g., mass and volume measurements) can be added, although absolute uncertainties with different units (±0.03 g and ±0.01 mL) cannot be combined. One could report the final result along with a relative uncertainty (e.g., density = 0.620 g/mL ± 6 parts per thousand). However, the more common practice is to multiply the relative uncertainty by the result to obtain an absolute uncertainty that has the same units as the result (0.620 g/mL ± 0.004 g/mL).

The rules for determining the number of significant figures in a derived result were set up so that the result would seem no less uncertain than the data from which it was derived. When actual uncertainties are computed, the number of digits known with absolute certainty may differ from the number predicted by the significant figures rules. In such a case, the computed uncertainties take precedence over the somewhat arbitrary significant figure rules. Thus, the volume considered in Example 4 should be reported as $7\times10^4\ \text{mm}^3 \pm 2\times10^4\ \text{mm}^3$.

Dealing With "Bad" Data

A question of concern to novice experimenters is "Must I report this measurement that totally disagrees with the others?" in a series of replicates (data obtained in a group of trials to determine a single value). Of course, any data that you know to be erroneous (such as a buret reading made when the stopcock was leaking or a mass of a sample that was spilled) should be rejected, with a brief notation made in the original data sheet explaining the reason for the deletion. But what if a measurement merely looks strange compared with other values? Then, you must use a mathematical test such as the one described below for guidance. In order to perform this test, you should identify the suspected bad datum and temporarily ignore it. Obtain the average of the remaining data. Compare each value used to obtain the average with the average itself and then calculate the average of these deviations. Next, compute the deviation of the suspect datum from the average of the other data. If the deviation of the suspect value from the average is more than four times the average deviation for the other data, the suspect value should be rejected. The use of this method is illustrated in the examples given below.

Example 1 The measured values are volumes of base used in a titration. The values are 25.19 mL, 26.25 mL, 26.01 mL, 26.17 mL, and 26.32 mL.

Suspect datum:	25.19 mL
Remaining data:	26.25 mL, 26.01 mL, 26.17 mL, and 26.32 mL
Average of remaining data:	26.19 mL
Deviations:	0.06 mL, 0.18 mL, 0.02 mL, and 0.13 mL
Average deviation:	0.098 mL
Deviation of suspect datum:	1.00 mL

1.00 > 4(0.098 mL): The suspect value is rejected and the average volume is reported as 26.19 mL.

Example 2 The values are 25.79 mL, 26.01 mL, and 26.17 mL.

Suspect datum:	25.79 mL
Remaining data:	26.01 mL and 26.17 mL
Average of remaining data:	26.09 mL
Deviations:	0.08 mL, 0.08 mL
Average deviation:	0.08 mL
Deviation of suspect datum:	0.30 mL

0.30 < 4(0.08): The suspect value is *not* rejected and the average volume is reported as 25.99 mL.

Getting Started in the Laboratory

EXP
1

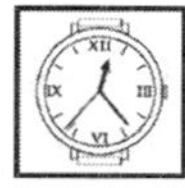

Laboratory Time Required

Three hours.

Special Equipment and Supplies

Analytical balance
Gas burner
Thermometer
Wooden applicator stick
Volumetric pipet, 10-mL
Ringstand with iron ring
Glass rod
Small paper labels
Buret
Buret clamp
Barometer
Candle (optional)
Ice

Objective

In the course of this experiment, the student will learn proper use of many common pieces of laboratory equipment, such as the burner, thermometer, pipet and buret.

Safety

The gas burner, hot glass, and boiling water used in the experiment can cause painful and serious burns to the skin. Hot glass looks like cold glass, so it pays to be cautious when heating or melting glass. You should also pay close attention when lighting or adjusting the burner and when measuring the temperature of boiling water.

Safety glasses with side shields are required.

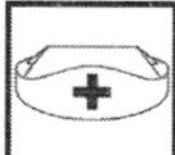

First Aid

Severe burns should be treated by a physician.

Preamble

The chemistry laboratory of motion pictures is an exciting but scary place—dimly lit; cluttered with smoking retorts; flasks filled with boiling, brightly colored liquid; and electrodes that discharge loudly for no apparent purpose. Although items such as these are still to be found in chemical museums or research labs, today's teaching labs tend to be furnished with more basic equipment and with less noise and smoke. However, the excitement is still there. It comes from observing an unfamiliar chemical transformation; making a set of measurements that clearly confirms a chemical law; or simply mastering a new laboratory skill.

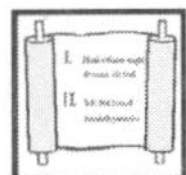

Principles

In this experiment, you will identify useful items of equipment and make your first measurements using four basic instruments: the balance, the pipet, the buret, and the thermometer. As will be the case for every experiment, you will be expected to follow the rules given in the Introduction pertaining to safety, chemical disposal, report preparation, significant figures, and the operation of instruments. If you are not familiar with these rules, reread the introductory section of this laboratory manual.

Our tour of the laboratory begins with an introduction to the desk equipment, which is used to contain, mix, measure, heat, stir, or separate chemicals, and for other purposes as well. Your kit of equipment will probably contain flasks and beakers of varying sizes, a wash bottle, a wire gauze, stirring rods, test tubes, and one or more graduated cylinders.

Your kit may also contain a gas burner of some sort. There is a variety of burners that are used in chemistry labs, all of which have means of containing a combustible mixture of gas and air. A mixture that has a high gas-to-air ratio will burn with a luminous yellow flame. Increasing the amount of air in the mix should produce a two-zone flame (Figure 1.1).

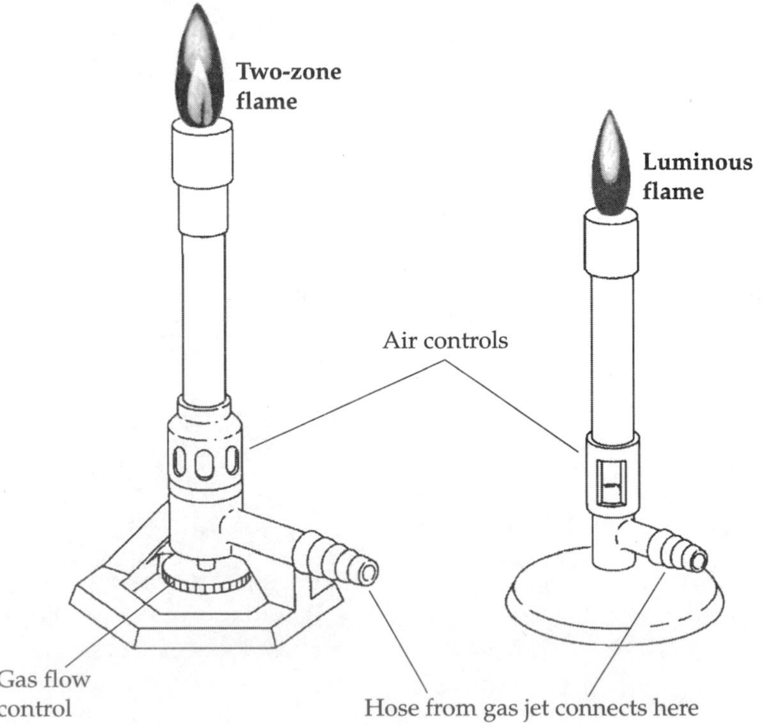

Figure 1.1 Gas burners

The two-zone flame is a more efficient flame than the yellow flame. The blue, inner zone of the flame is the cooler of the zones. The hotter, outer zone is relatively colorless and transparent. If a glass rod is placed in the outer zone, a yellow flame may be observed above the rod. This flame has the characteristic appearance of thermally excited sodium atoms. (Sodium is a component of glass.)

The outer zone is hot enough to soften "soft" glass so that it can be bent, stretched, blown, or fire polished to fabricate custom laboratory glassware. Borosilicate glasses, such as Pyrex®, however, have higher melting points and cannot be worked in this way, although you can still use the Bunsen flame to round sharp edges slightly. Figure 1.2 shows soft glass and Pyrex rods whose ends have been rounded or "fire polished."

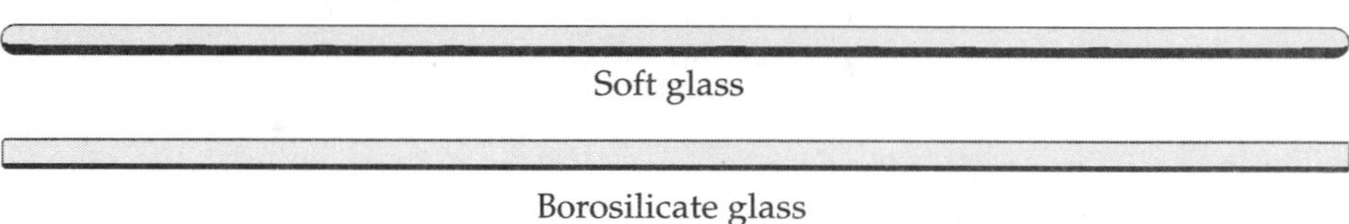

Figure 1.2 Fire-polished glass rods

Although it is unlikely to be a part of your laboratory kit, the thermometer is an instrument that you will use frequently in the course of your lab work. The thermometer is used to measure the temperature of an object on the Fahrenheit (°F), Celsius (°C), or Kelvin (K) scale. The latter two are most useful to chemists, and are easily interconverted using the formula shown in Equation 1.1.

$$t_{\text{Celsius}} = T_{\text{Kelvin}} - 273.16 \qquad (1.1)$$

The Celsius thermometer is easily calibrated by comparing its readings with the normal freezing point and normal boiling point of water, which are exactly 0°C and exactly 100°C, respectively, at an atmospheric pressure of 760 torr. Because the boiling point of water varies slightly with pressure, you will have to take this into account. To make this correction, you must know the barometric pressure in the laboratory. You may then look up the actual boiling point of water at that pressure in the *Handbook of Chemistry and Physics,* or, if you wish, you can calculate a value for the boiling point using the expression given in Equation 1.2, where P is the barometric pressure in torr and T_b is the boiling point of water on the Kelvin scale.

$$T_b = \frac{2124}{8.573 - \log P} \qquad (1.2)$$

Another instrument you will frequently use in the general chemistry laboratory is the balance. You will probably need to be familiar with only two types of balance: a top-loading balance for weighings in which a tolerance of ±0.1 g is acceptable, and an analytical balance for weighings that require a tolerance of ±0.001 g or smaller. General instructions concerning the use of balances are given in the Introduction. Your instructor will give you more specific directions for using the type of balance found in your lab.

In this experiment, you will use a top-loading or analytical balance to weigh samples of water that have been dispensed by a pipet and a buret. The data on the masses of the water samples will be used to **calibrate** the pipet and buret (determine, accurately, what volume of water the pipet and buret actually dispensed).

The volumetric pipet (see Figures I.3 and I.4 in the Introduction) is used to dispense a known volume of liquid quickly and precisely. A clean 10-mL pipet is probably precise (or reproducible), but may not be completely accurate because a small tolerance in volume dispensed is deemed acceptable.

Table 1.1 Specifications of 10-mL Pipets

Type	*Tolerance*	*Cost*
Class A	± 0.02 mL	\$6–\$10
Class B	± 0.04 mL	\$4–\$5
Polypropylene	± 0.06 mL	\$16

For three grades of 10-mL pipet, the tolerances and approximate cost per pipet are shown in Table 1.1. Although the polypropylene pipet has a higher cost and poorer precision than the glass pipets, it has one great advantage: it is unbreakable.

Like the pipet, the buret (see Figure I.3) is used to dispense liquid samples of known volumes. Burets are engraved with a scale reading downward, from 0 mL to 50 mL, and measure *volume dispensed*. A buret would normally be used when a variable amount of liquid is needed in an experiment, as in a titration. The buret used in introductory laboratory courses is usually accurate to ±0.03 mL.

You can improve the accuracy of experiments that employ pipets or burets by calibrating the instruments. This is done by weighing samples of water dispensed by the pipet or by the buret and recording the water temperature. You can then use the mass and temperature data to calculate the volume of the water dispensed by the pipet or by the buret by making use of one of water's fundamental properties – its density.

Density is defined as the ratio of an object's mass to its volume (see Equation 1.3). Densities for solids and liquids typically have units of g/mL.

$$\text{Density} = \frac{\text{mass}}{\text{volume}} \tag{1.3}$$

The density of water depends, to some extent, on its temperature. A table of the density of water at various temperatures is found in Appendix B.

Once you have determined the mass of the water delivered by your pipet or buret and looked up its density, you will use Equation 1.3 to calculate the actual volume of water dispensed in each of your calibration trials. You will use your results in answering several questions about the suitability of various pieces of glassware in different experimental settings.

Procedure

Desk Equipment

Using Figures I.1 through I.3 in the Introduction as a guide, try to identify each item that is on your equipment list. Note the names of unfamiliar items so that you will recognize them when they are called for in subsequent experiments. If any items are missing, broken, or inoperable, follow your instructor's directions to obtain replacements.

Procedure in a Nutshell

Examine your desk equipment. Adjust the burner to give a yellow flame and then to give a blue flame with a well-defined set of two cones. Calibrate your thermometer with ice water and with boiling water. Calibrate your pipet and buret by delivering water to tared flasks. Answer post-laboratory questions concerning the use of glassware in the laboratory.

The Gas Burner

Connect your gas burner (see Figure 1.1) to the gas source using rubber tubing and check the operation of the air and gas controls before lighting the burner. If you are using a match to light your burner, set the air control to partially open, light the match first, gradually turn on the gas, and cautiously light the burner. If you are using a flint igniter, set the air control to partially open, gradually turn on the gas, and cautiously light the burner. Note the effect of changing the gas flow, and of opening and closing the air control.

Adjust the burner to obtain a luminous yellow flame, and hold a large beaker so that its bottom is briefly in the outer zone, and note whether a deposit forms or not. Answer the questions on the Summary Report Sheet.

You may wish to compare the results of this to the results of holding a large beaker over the flame of a candle, which burns with much less intensity than a burner. If you do, answer the questions on the Summary Report Sheet.

Adjust the burner to give a two-zone flame, as in Figure 1.1. Place a wooden applicator stick so that it passes through the inner zone as well as the outer zone. Note your observations on the Summary Report Sheet.

Caution!

DO NOT TOUCH THE HOT GLASS!
IT LOOKS THE SAME AS COOL GLASS.

The Thermometer

Fill a 100-mL beaker about half full with ice and add just enough water to cover the ice. Stir the mixture gently with the thermometer at intervals until a constant reading is observed. Record this melting point temperature (to ±0.1°C) on the Report Sheet. (We use the terms *melting point* and *freezing point* interchangeably; they are the same.)

Fill a 250-mL beaker two-thirds full with water, add a few boiling chips, and heat the water to boiling over a Bunsen burner. The proper set up is shown in Figure 1.3. When the water is boiling smoothly, immerse the thermometer to the immersion mark (if possible) and cautiously read the boiling temperature to ±0.1°C. Repeat a few minutes later to verify the accuracy of your measurement. Avoid touching the bottom of the beaker with the thermometer bulb.

Label your thermometer with the freezing point and boiling point that you have found by using it. If your freezing point and boiling point readings are too low or too high by an identical amount, you may use that amount as your thermometer correction. If the two deviations are not identical, you should plot the observed freezing and boiling temperatures versus the theoretical values for the freezing point and boiling point, and draw a straight line through the experimental points. This graph will permit you to report the corrected temperature for any thermometer reading.

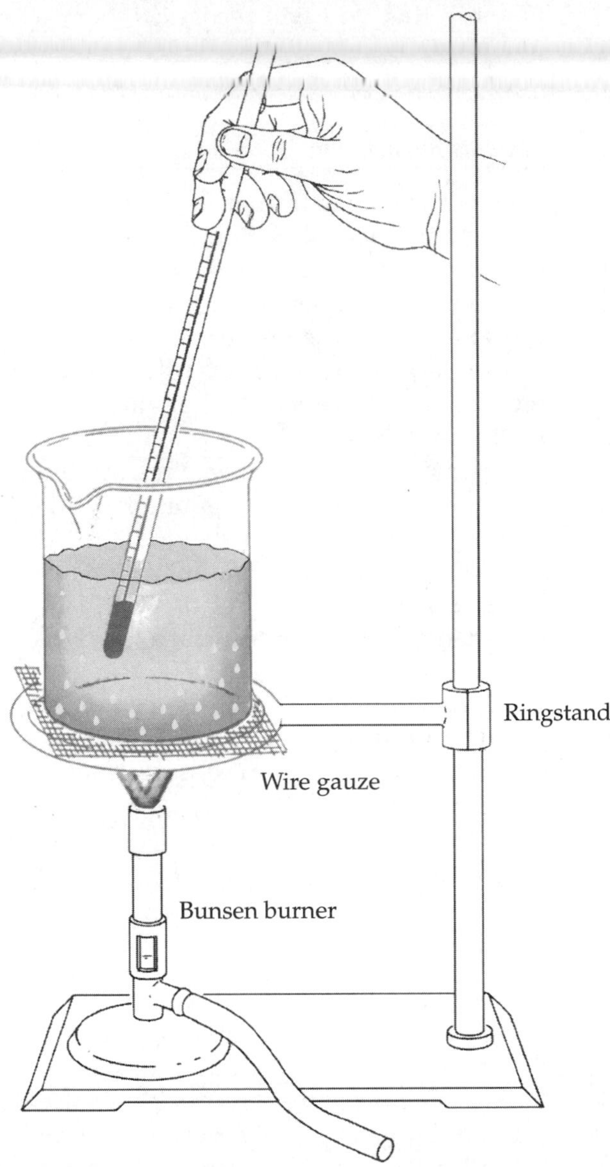

Figure 1.3 Boiling water

The Pipet

Clean and dry a small Erlenmeyer flask and close the flask with a cork or rubber stopper of appropriate size. Take the stoppered flask to your assigned balance and determine its mass, following the weighing instructions given in the Introduction.

Practice filling your pipet using the pipet bulb until you have mastered this important skill. The steps are described in the Introduction and are illustrated in Figure I.4.

When you are satisfied with your technique, use the pipet to dispense a 10-mL volume of water into the previously weighed flask. Replace the stopper and reweigh the flask, using the same balance. Record the mass on your Report Sheet. Dispense and weigh a second volume of water using the same procedure. Repeat this procedure another time, so that you will have delivered a total of three 10-mL portions of water into the flask. Before discarding the excess water, measure its temperature using your calibrated thermometer. Record the corrected temperature.

Use Equation 1.3 and your data to find the volumes of the water samples delivered by your pipet. Report the average of these values, along with the average deviation of the individual measurements from this average value.

The Buret

Prepare and weigh a small Erlenmeyer flask as you did when you were calibrating the pipet. Clean a buret, following the steps described in the Introduction. Fill the buret with distilled water; the water level should be just below the 0-mL mark. Record the initial buret reading. (The procedure for reading the buret is described in Figure I.5; the procedure for reading the meniscus is shown in Figure 1.4.) Use the buret to dispense a 10-mL volume of water into the previously weighed flask. Record the final buret reading. Replace the stopper and reweigh the flask, using the same balance. Record the mass on your Report Sheet. Dispense and weigh an additional 10-mL volume of water using the same procedure, and without emptying the Erlenmeyer flask. Repeat the procedure another time so that you will have obtained the masses of three separate 10-mL volumes of water dispensed by the buret.

Use Equation 1.3 and your data to find the actual volumes of the three water samples delivered by your buret. Report the average of these values, along with the average deviation of the individual measurements from this average value.

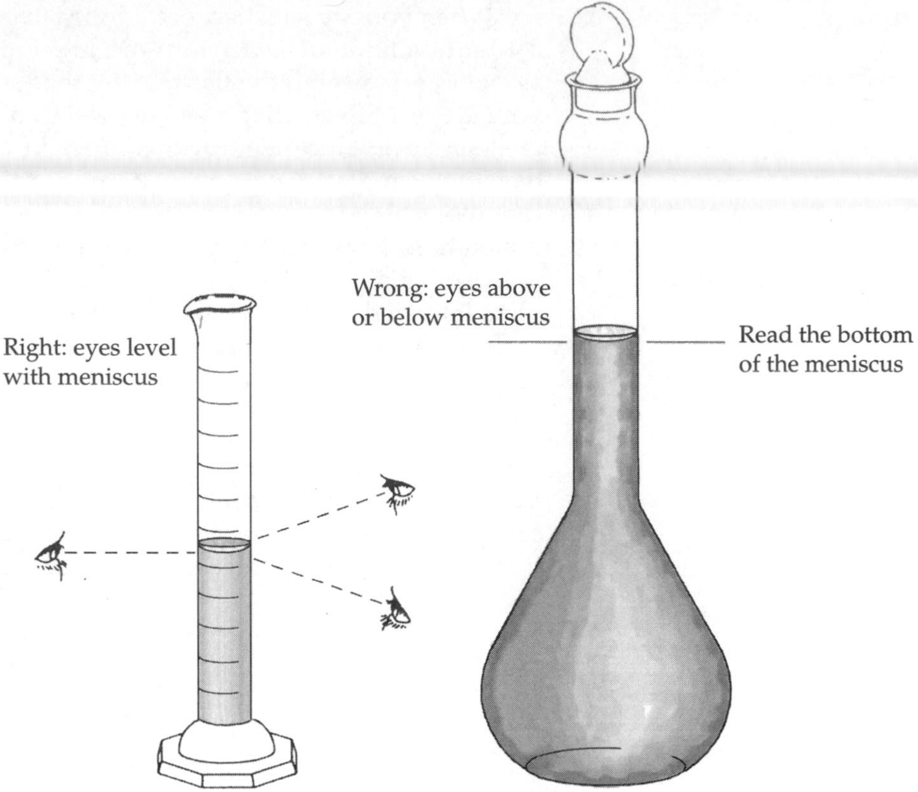

Figure 1.4 Reading the meniscus in volumetric glassware

Disposal of Reagents

All chemicals used in this experiment may be diluted with water and flushed down the sink.

Questions

1. Using your data for the freezing point and boiling point of water, plot the actual Celsius temperature versus the observed (thermometer) temperature and draw a straight line through the points. Suppose your thermometer indicates that a liquid boils at 65.3°C. What would be its actual boiling point?

2. Why should you always be careful to use the same balance throughout an experiment?

3. In the colorimetric analysis of samples for phosphate content, it is necessary to add 10.00 mL of a color-enhancing reagent to each of several samples. Would it be preferable to use a pipet or a buret in dispensing the reagent? Explain your answer briefly.

4. In the titration of vinegar samples, one needs to know the exact volume (within one drop (0.05 mL)) of basic solution that must be added to each sample in order to cause an indicator to change color. Would it be preferable to use a pipet or a buret in dispensing the basic solution? Explain your answer briefly.

name | section | date

Pre-Lab Exercises for Experiment 1

These exercises are to be completed after you have read the experiment but before you come to the laboratory to perform it.

1. Without looking at the figures, sketch the laboratory equipment specified.

beaker	pipet
Erlenmeyer flask	spatula

2. A thermometer indicates that water freezes at 0.5°C and boils at 100.2°C at a barometric pressure of 760 torr. The observed freezing point of an unknown liquid is –1.3°C.

a. What is the corrected freezing point of the unknown, in °C?

b. What is this freezing point on the Kelvin scale?

3. Suppose a 10-mL pipet dispenses 9.99 g of water at 22°C. What volume of water was dispensed by the pipet? Show your calculations. (You can find the density of water at 22°C in Table B.1 at the back of this manual). Is the pipet within tolerance limits for a Class A pipet? For a Class B pipet?

4. Suppose that you dispense water from a buret into a flask as described in this experiment. The initial and final buret readings are 0.05 mL and 10.46 mL, respectively, and the flask is observed to weigh 48.0017 g when empty, and 56.8101 g after the water sample has been dispensed into it. The temperature of the water was 22°C. Calculate:
   a. the apparent volume of the water dispensed;

   b. the actual volume of the water dispensed;

   c. the % error.

name ____________________ section ____________________ date __________

Summary Report on Experiment 1

The Gas Burner

1. The yellow luminous flame
   a. What deposits on the bottom of the beaker that is held in the flame?

   b. What is the origin of this deposit?

   c. How is this deposit related to the yellow flame color?

   d. *Optional*: Candle Flame: Answer the same three questions but concern yourself with the candle flame.

2. The two-zone flame
   a. In which zone does the wooden applicator burn first?

   b. What do you conclude from this observation?

Thermometer Calibration

Observed melting point of ice ____________________

Thermometer correction ____________________

Observed boiling point of water ____________________

Barometric pressure ____________________

Actual boiling point of water ____________________

Thermometer correction ____________________

Pipet Calibration

	Trial 1	*Trial 2*	*Trial 3*
Mass of stoppered flask plus 10 mL water			
Mass of empty stoppered flask			
Mass of water dispensed			
Thermometer reading			
Corrected temperature of water			
Density of water at observed temperature			

Calculated volume of water (show calculations):

	Trial 1	*Trial 2*	*Trial 3*
Calculated volume			

Average volume ______________________ ± Average deviation ______________________ name ______________ section ______________ date ______________

Buret Calibration

	Trial 1	*Trial 2*	*Trial 3*
Mass of stoppered flask plus 10 mL water			
Mass of empty stoppered flask			
Mass of water dispensed			
Thermometer reading			
Corrected temperature of water			
Density of water at observed temperature			

Calculated volume of water (show calculations):

	Trial 1	*Trial 2*	*Trial 3*
Calculated volume			

Average volume ______________ ± Average deviation ______________

Freezing and Melting of Water[1]

EXP
1P

Preamble

Freezing temperature, the temperature at which a substance turns from liquid to solid, and melting temperature, the temperature at which a substance turns from a solid to a liquid, are characteristic physical properties. In this experiment, the cooling and warming behavior of a familiar substance, water, will be investigated. By examining graphs of the data, the freezing and melting temperatures of water will be determined and compared.

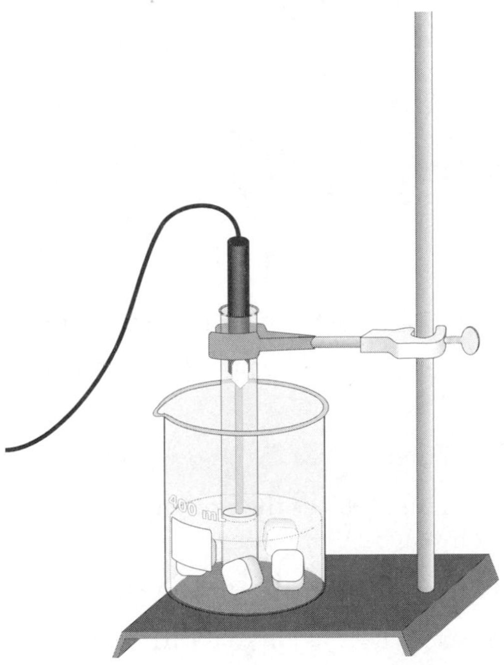

Figure 1P.1

Materials

CBL 2 interface	400-mL beaker
TI Graphing Calculator	10-mL graduated cylinder
DataMate program	Test tube
Temperature probe	Salt
Ringstand	Ice
Utility clamp	Water

[1] Adapted from Holmquist, D.D., J. Randall, D.L. Votz, *Chemistry with Calculators: Chemistry Experiments Using Vernier Sensors with Texas Instruments CBL 2™* (Vernier Software and Technology, Beaverton, Oregon: 2000). Used with permission.

Procedure

Part I Freezing

1. Put about 100 mL of water and 6 ice cubes into a 400-mL beaker.

2. Put 5 mL of water into a test tube and use a utility clamp to fasten the test tube to a ringstand. The test tube should be situated above the water bath. Place the temperature probe into the water inside the test tube.

3. Plug the temperature probe into Channel 1 of the CBL 2 interface. Use the link cable to connect the TI Graphing Calculator to the interface. Firmly press in the cable ends.

4. Turn on the calculator and start the DataMate program. Press [CLEAR] to reset the program.

5. Set up the calculator and interface for the temperature probe.

a. Select SETUP from the main screen.

b. If the calculator displays a temperature probe in CH 1, proceed directly to Step 6. If it does not, continue with this step to set up your sensor manually.

c. Press [ENTER] to select CH 1.

d. Select TEMPERATURE from the SELECT SENSOR menu.

e. Select the temperature probe you are using (in °C) from the TEMPERATURE menu.

6. Set up the data-collection mode.

a. To select MODE, press [▲] once and press [ENTER].

b. Select TIME GRAPH from the SELECT MODE menu.

c. Select CHANGE TIME SETTINGS from the TIME GRAPH SETTINGS menu.

d. Enter "10" as the time between samples in seconds.

e. Enter "90" as the number of samples. The length of the data collection will be 15 minutes.

f. Select OK to return to the setup screen.

g. Select OK again to return to the main screen.

7. When everything is ready, select START to begin collecting data. Lower the test tube into the ice-water bath.

8. Soon after lowering the test tube, add 5 spoons of salt to the beaker and stir with a stirring rod. Continue to stir the ice-water bath throughout the remainder of Part I.

9. Slightly, but continuously, move the temperature probe during the first 10 minutes of Part I. Be careful to keep the probe in, and not above, the ice as it forms. When 10 minutes have gone by, stop moving the probe and allow it to freeze into the ice. Add more ice cubes to the beaker as the original ice cubes get smaller.

10. Data collection will stop after 15 minutes. Keep the test tube *submerged* in the ice-water bath until Step 13.

11. Analyze the flat part of the graph to determine the freezing temperature of water. To do this:

a. Press ENTER to return to the main screen, then select ANALYZE.

b. Select STATISTICS from the ANALYZE OPTIONS menu.

c. Use ▶ to move the cursor to the beginning of the flat section of the curve. Press ENTER to select the left boundary of the flat section.

d. Move the cursor to the end of the flat section of the graph, and press ENTER to select the right boundary of the flat section. The program will now calculate and display the statistics for the data between the two boundaries.

e. Record the MEAN value as the freezing temperature in your data table. (Round to the nearest 0.1°C.)

f. Press ENTER to return to the ANALYZE OPTIONS menu, then select RETURN TO MAIN SCREEN.

12. Store the data from the first run so that it can be used later. To do this:

a. Select TOOLS from the main screen.

b. Select STORE LATEST RUN from the TOOLS MENU.

Part II Melting

13. Choose START to begin data collection, then raise the test tube and fasten it in a position above the ice-water bath. Do not move the temperature probe during Part II.

14. Dispose of the ice water by pouring it down the drain. Obtain 250 mL of warm tap water in the beaker. When 12 minutes have passed, lower the test tube and its contents into this warm-water bath.

15. Data collection will stop after 15 minutes. Analyze the flat part of the graph to determine the melting temperature of water. To do this:

a. Press ENTER to return to the main screen, then select ANALYZE.

b. Select STATISTICS from the ANALYZE OPTIONS menu.

c. Use ▶ to move the cursor to the beginning of the flat section of the curve. Press ENTER to select the left boundary of the flat section.

d. Move the cursor to the end of the flat section of the graph, and press ENTER to select the right boundary of the flat section. The program will now calculate and display the statistics for the data between the two boundaries.

e. Record the MEAN value as the freezing temperature in your data table. (Round to the nearest 0.1°C.)

f. Press ENTER to return to the ANALYZE OPTIONS menu, then select RETURN TO MAIN SCREEN.

16. A good way to compare the freezing and melting curves is to view both sets of data on one graph.

a. Select GRAPH from the main screen, then press ENTER.

b. Select MORE, then select L2 AND L3 VS L1 from the MORE GRAPHS menu.

c. Both temperature runs should now be displayed on the same graph. Each point of Part I (freezing) is plotted with a box, and each point of Part II (melting) is plotted with a dot.

17. Print a graph of temperature *vs.* time (with two curves displayed). Label each curve as "freezing of water" or "melting of ice."

Data Table

Freezing temperature of water	°C
Melting temperature of water	°C

name section date

Processing the Data

1. What happened to the water temperature during freezing? During melting?

2. According to your data and graph, what is the freezing temperature of water? The melting temperature? Express your answers to the nearest 0.1°C.

3. How does the freezing temperature of water compare to its melting temperature?

4. Tell if the *kinetic energy* of the water in the test tube increases, decreases, or remains the same in each of these time segments during the experiment when:
   a. the temperature is changing at the beginning and end of Part I

   b. the temperature remains constant in Part I

   c. the temperature is changing at the beginning and end of Part II

   d. the temperature remains constant in Part II

5. In those parts of Question 4 in which there was no kinetic energy change, tell if *potential energy* increased or decreased.

On the Nature of Pennies

EXP
2

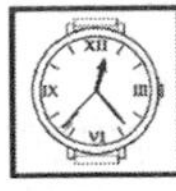

Laboratory Time Required

Three hours.

Special Equipment and Supplies

- Balance
- Burner
- Tongs
- Crucible
- Triangle
- Ringstand
- Iron ring
- Coffee cup calorimeter
- Thermometer
- Graduated cylinder, 100-mL
- Calipers
- Ruler
- Iron file
- Magnet
- Pennies
- Vinegar, $HC_2H_3O_2(aq)$[1]
- Table salt, NaCl(*s*)
- Hydrochloric acid, HCl(*aq*)

Objective

In the course of this experiment, the student will formulate and test hypotheses meant to explain observed variations in the masses of one-cent coins.

Safety

The gas burner can cause painful and serious burns to the skin. Be careful when lighting or adjusting the gas burner and when handling hot metals. Do not heat metals over an open flame. Use a crucible.

Hydrochloric acid is corrosive and may cause skin burns or irritation.

Safety glasses with side shields are required.

First Aid

Following skin contact with hydrochloric acid, rinse the affected area thoroughly with water.

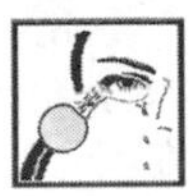

If hydrochloric acid enters the eyes, rinse them in the eyewash (at least 20 minutes of flushing with water is recommended) and seek medical treatment.

Severe burns should be treated by a physician.

---

[1] Vinegar is an aqueous solution of acetic acid. If commercial vinegar is not available, 1 M acetic acid may be substituted for it. The 1 M acetic acid may be labeled with its name or a variation of its formula, most commonly $HC_2H_3O_2$ or CH_3CO_2H.

Preamble

One-cent pieces currently in circulation exhibit a number of different masses. In this experiment, students are asked to determine whether or not variations in the mass of a penny are significant and to develop and test hypotheses that explain both the major and minor variations.

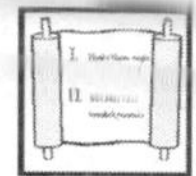

Principles

Hundreds of years ago, the value of a coin was determined by its mass and the worth of the metal it contained. It was common for merchants to weigh coins before making change to ensure that valuable metal had not been "shaved" from the coins. Today, the value of our currency is determined by the pledge of the federal government. Thus, damaged coins may remain in circulation, as long as their denomination is recognizable.

Coins are frequently made of silver, gold, or copper. These coinage metals have low chemical activity[2], with good reason. You would not want to burn your hands when handling money, as would happen if coins were made of sodium and potassium, and you would not want your money to crumble, as rusting iron coins would. A common test of metal activity involves treating metal samples with hydrochloric acid. (Although hydrogen is not a metal, it does lose electrons when it combines with nonmetals to form acids.) Metals that are more active than hydrogen will be oxidized (i.e., will lose electrons) causing the hydrogen ions in the acid to be reduced to hydrogen gas, which will bubble out of solution. Metals that exhibit this behavior when treated with hydrochloric acid are said to be "replacing" hydrogen in the acid. Neither gold nor silver nor copper is active enough to replace hydrogen from acid. In contrast, hydrogen gas bubbles out of solution when more active metals—such as iron or zinc—are treated with dilute hydrochloric acid. Sodium and potassium are active enough to replace hydrogen from water.

It is sometimes desirable to use an active metal in situations where its reactivity might be a problem. Iron is used in construction because of its strength. However, the iron is often alloyed with other elements, producing steel, which is more resistant to corrosion than is pure iron. Chromium and tin are often used to plate steel because they oxidize to form durable oxide coatings that prevent the interaction of oxygen and the iron beneath the surface layer.

Although copper is not a very active metal, it too becomes coated when exposed to air. Copper metal exposed to the atmosphere for a long period of time changes from its bright reddish-brown metallic luster to an almost black color and then may go on to develop a green patina. The initial blackening results from the formation of copper oxide and copper sulfide. These salts can be removed from the surface of pennies by placing the coins in vinegar to which a bit of table salt has been added. Interestingly, leaving the pennies soaking in vinegar overnight results in the formation of green deposits on the surfaces of the coins.

Neither the mass nor the volume of a metal sample is a characteristic of the metal itself. However, the ratio of the mass to the volume (the density) is a characteristic of the metal. When attempting to determine the identity of a metal, one frequently weighs a sample (to determine its mass) and then finds the volume by direct measurement of dimensions (if the sample has a regular shape) or by measuring the volume of water dis-

[2] This means the metals are not easily oxidized. It also means that it is relatively easy to decompose compounds of the metal so that the free metal can be obtained in element form.

placed when the sample is submerged. The unit generally used for density is g/mL, which is equivalent to g/cm^3, g/cc, and kg/dm^3.

Other characteristics that might be used to identify metals are specific heat, melting point, and appearance. The specific heat of a substance is a measure of the amount of energy needed to raise the temperature of one gram of the substance by one degree Celsius. A typical unit for specific heat is J/g·°C; this is the equivalent of J/g·K because changes in temperature have the same magnitude on the Celsius and Kelvin scales. One method of determining the specific heat of metal involves placing a weighed sample of the metal in a bath of boiling water and allowing the metal to be warmed to the temperature of the boiling water. The metal sample is then transferred quickly to an insulated container, which holds a sample of cool water for which the mass and temperature are known. The specific heat (C_M) of the metal can be calculated from the relation shown below, where g_M and g_{H_2O} represent the mass of the metal and the mass of the cool water, respectively, and t_b, t_c, and t_f represent the temperatures of the boiling water, the cool water, and the final system (metal plus cool water), respectively. The specific heat of water (C_{H_2O}) has a value of 4.18 J/g °C. The relationship between these variables is shown in Equations 2.1 and 2.2.

$$\text{Energy Change for Water} = -(\text{Energy Change for Metal}) \quad (2.1)$$

$$g_{H_2O}C_{H_2O}\,(t_f - t_c) = -g_M\,C_M\,(t_f - t_b) \quad (2.2)$$

The melting point of a substance is the temperature at which the solid and liquid forms of the substance are in equilibrium. A temperature of approximately 650°C can be achieved with a typical laboratory burner. Therefore, metals with melting points of 650°C or below will melt when held in a burner flame.

The appearance of a material might be characterized by such properties as color, shininess, and type of crystal structure. Table 2.1 lists several metals and their densities, specific heats, and melting points, along with some information about each metal's appearance.

Table 2.1 Characteristics of Some Metals

	Density at 20°C, g/mL	*Specific Heat, J/g·°C*	*Melting Point, °C*	*Appearance*
Aluminum	2.70	0.89	660	Silvery white, soft
Zinc	7.13	0.39	420	Bluish-white, lustrous
Iron	7.87	0.45	1535	Hard, brittle
Copper	8.96	0.39	1083	Reddish, lustrous
Silver	10.5	0.235	962	Brilliant white, lustrous
Lead	11.34	0.13	328	Bluish-white, very soft
Gold	19.32	0.13	1064	Yellow, soft

Procedure

Procedure in a Nutshell

Obtain the masses of a number of pennies and decide how many sets of pennies are present in the collection studied. Propose hypotheses to explain the differences in mass among the various sets. Perform experiments to test your hypotheses and state whether the experiments confirm or debunk the hypothesis being tested. Make suggestions for further study if appropriate.

Obtain a collection of pennies from your instructor. Weigh the pennies individually on the analytical balance. Record the data called for (mass of penny, date of mintage, description) on the Summary Report Sheet. Determine how many sets of pennies you have in your collection. Members of a single set of pennies may exhibit small variations about an average mass, which should be significantly different from the average mass of pennies in any other set.

Consider each of the hypotheses listed in your Pre-Lab Exercises to explain why the pennies do not all have the same mass. Decide whether your hypotheses are best suited to explaining the minor variations of the masses of the pennies in a single set or the major variation between the different sets. Analyze your initial data to determine if they are sufficient to confirm or refute each hypothesis. If they are not sufficient in a given case, develop and perform additional experiments aimed at confirming or refuting that hypothesis. Describe the experiments and record the data obtained from them on the appropriate spaces in the Summary Report Sheet.

Your discussion should consider how successfully your hypotheses explained the variations in mass. Estimate uncertainties in your numerical results and discuss possible improvements to your experimental design and/or further experimentation, which might be needed.

name section date

Pre-Lab Exercises for Experiment 2

These exercises are to be completed after you have read the experiment but before you come to the laboratory to perform it.

1. List five hypotheses (to explain the difference of the pennies' masses) either suggested in the Principles section, or developed from some other source. State whether the hypothesis (if correct) would explain major or minor variations of the masses of the pennies.

2. Explain briefly how you might refute or obtain support for each of the hypotheses listed above.

name section date

Summary Report on Experiment 2

Mass of Penny	*Mintage Date*	*Description of Penny*	*Set*

Hypothesis: ______________________________

____

____

____

Would explain: ☐ major ☐ minor variation

Further Experimentation

Procedure: ______________________________

____

____

____

____

Data: ______________________________

____

____

____

____

Conclusions: ______________________________

____

____

____

____

name section date

Hypothesis:

Would explain: ☐ major ☐ minor variation

Further Experimentation

Procedure:

Data:

Conclusions:

Hypothesis: ____

____

____

____

Would explain: ☐ major ☐ minor variation

Further Experimentation

Procedure: ____

____

____

____

____

Data: ____

____

____

____

____

Conclusions: ____

____

____

____

____

name section date

Hypothesis:

Would explain: ☐ major ☐ minor variation

Further Experimentation

Procedure:

Data:

Conclusions:

Hypothesis: ______________________________

____

____

____

Would explain: ☐ major ☐ minor variation

Further Experimentation

Procedure: ______________________________

____

____

____

____

Data: ______________________________

____

____

____

____

Conclusions: ______________________________

____

____

____

____

Density

EXP
3

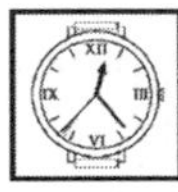

Laboratory Time Required

Two hours.

Special Equipment and Supplies

Analytical balance
Graduated cylinders, 10 mL and 25 mL
Vernier calipers
Metric ruler or meter stick
Volumetric pipet, 10 mL
Pipet bulb
Buret
Stirring rod
Evaporating dish
Hot plate
Cylinders of aluminum, Al(*s*)
Sand, SiO_2 (*s*)
Table salt, NaCl(*s*)
Water, H_2O (*l*)
Isopropyl alcohol, $(CH_3)_2CHOH$ (*l*)

Objective

In performing this experiment, the student will learn methods for determining the density of several substances and will be given the opportunity to consider how much "empty space" is encompassed in a "solid" sample.

Safety

Although aluminum is not hazardous, it should not be exposed to acids with which it might react. Aluminum can react with acids, liberating explosive hydrogen gas.

Replace any broken glassware to avoid cuts.

Safety glasses with side shields are required.

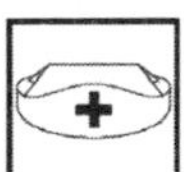

First Aid

For serious cuts, apply direct pressure if necessary to control bleeding and summon medical help.

Preamble

Weighing, using the analytical balance, and measuring volumes of liquids, using volumetric glassware, are two of the most fundamental operations performed by a chemist. Because any quantitative experiment will most likely require one or both of these operations, it is important that you become proficient in weighing and measuring early in your laboratory training. In this experiment you will measure both the mass and the volume of several samples and use various methods to determine the amount of "empty" space that each sample encloses.

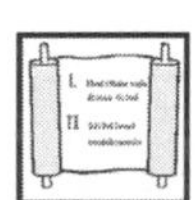

Principles

The density of a substance is defined as its ratio of mass to volume. Consequently, density has such dimensions as lb/ft^3 or g/cm^3 (which is the same as g/mL). The density of an element or compound is a characteristic physical property at a specified temperature and atmospheric pressure and is therefore useful in identifying the substance.

In the general chemistry laboratory, density is most often determined by finding both the mass and volume of a substance, independently, and calculating the density by dividing the mass by the volume. Mass determinations are easily accomplished by weighing a sample on a balance. However, the determination of volume is not always straightforward and methods may vary in accord with the nature of the sample. For instance, if the solid is a large object of regular shape, such as cubic or cylindrical, one can measure its dimensions and calculate the volume from the appropriate formula ($V = s^3$ for a cube and $V = \pi r^2 h$ for a cylinder). If the object is large, but of irregular shape, one can submerge it in water and determine the volume of the water displaced (according to Archimedes' Principle, the volume of the liquid displaced is equal to the volume of the object that does the displacing). If the sample is composed of small particles, as sand and table salt are, one might be tempted to fill a graduated cylinder with sample and read the volume from the calibration marks. This could lead to a large error in the determination of the volume because such granular samples have a considerable amount of empty space between particles. A more accurate determination of the volume of such samples may be obtained by adding a known volume of liquid to the sample, stirring to ensure mixing of the solid and the liquid, and reading the total volume of the mixture from the calibration marks. The actual volume occupied by the particles of granular solid is then determined as the difference between the total volume (of the liquid plus solid) and the volume of the liquid alone. This method is effective as long as the solid does not dissolve in the liquid.

Procedure

Weigh an aluminum cylinder to the nearest 0.1mg, using the analytical balance, and record the mass on the Summary Report Sheet. Determine the dimensions of the cylinder using Vernier calipers and/or a ruler. Enter your data on the Summary Report Sheet. Place the aluminum cylinder in a 25-mL graduated cylinder. Clean a buret and fill it with deionized water. Make sure the water meniscus is at, or below, the top (0.00 mL) calibration mark. Use the buret to deliver 10.00 mL of water to the 25-mL graduated cylinder into which the aluminum cylinder has been placed. Record the volume of the (aluminum plus water) in the graduated cylinder.

Procedure in a Nutshell

Determine the densities of an aluminum cylinder, a sample of sand, and a sample of salt from independent measurements of each substance's mass and volume. Mix the sand sample and the salt samples with liquids to determine the actual volume occupied by the particles in these substances. Use data on the atomic radius of aluminum to determine the actual volume of the atoms in the cylinder.

Weigh another clean, dry 25-mL graduated cylinder to the nearest 0.1 mg, using the analytical balance, and record its mass on the Summary Report Sheet. Fill this second 25-mL graduated cylinder with sand up to the 10-mL mark. Weigh the sand-filled cylinder on the same balance used to determine the mass of the empty cylinder and enter your data on the Summary Report Sheet. Deliver 10.00 mL of water from the buret into the sand-filled graduated cylinder. Stir the mixture to ensure that the water penetrates to the bottom of the column of sand. Record the volume of the sand-and-water mixture on the Summary Report Sheet.

Weigh a third clean, dry 25-mL graduated cylinder to the nearest 0.1 mg, using the analytical balance, and record its mass on the Summary Report Sheet. Fill the 25-mL graduated cylinder with table salt up to the 10-mL mark. Weigh the salt-filled cylinder on the same balance used to determine the mass of the empty cylinder and enter your data on the Summary Report Sheet. Deliver 10.00 mL of water from the buret into the salt-filled graduated cylinder. Stir the mixture to ensure that the water penetrates to the bottom of the column of salt. Record the volume of the salt-and-water mixture on the Summary Report Sheet. Transfer the salt-water mixture to an evaporating dish and place the dish on a hot plate. Heat the evaporating dish gently until all of the water has evaporated. Determine the mass of the salt left in the evaporating dish. Record the mass on the Summary Report Sheet.

Weigh a fourth clean, dry 25-mL graduated cylinder to the nearest 0.1 mg, using the analytical balance, and record its mass on the Summary Report Sheet. Fill this fourth 25-mL graduated cylinder with salt up to the 10-mL mark. Weigh the salt-filled cylinder on the same balance used to determine the mass of the empty cylinder and enter your data on the Summary Report Sheet. Use a pipet to deliver 10.00 mL of isopropyl alcohol into the salt-filled graduated cylinder. Stir the mixture to ensure that the alcohol penetrates to the bottom of the column of salt. Record the volume of the salt-and-alcohol mixture on the Summary Report Sheet.

Disposal of Reagents

Return the aluminum cylinders to the reagent cart. Wet sand can be placed in the containers for solid waste. Mixtures of salt and water and salt and alcohol can be flushed down the drain.

Questions

1. Although the aluminum cylinder is a solid (and, therefore, is quite incompressible), it actually contains quite a bit of empty space. Consider an aluminum atom to be a sphere with radius equal to 143 pm and calculate the volume actually occupied by the atoms in your aluminum cylinder ($1\text{ pm} = 10^{-12}\text{ m}$; $V_{\text{sphere}} = \frac{4}{3}\pi r^3$).

2. Discuss the advantages and disadvantages of using the buret, rather than the pipet, to deliver 10.00 mL of liquid.

3. Calculate your percentage yield of salt in the evaporation process.

$$\text{Percentage yield} = \frac{\text{mass of salt recovered}}{\text{mass of salt dissolved}} \times 100\%$$

4. You reported two densities for each of the materials studied (aluminum, sand, and table salt). For each material, state which density is more accurate. Explain your choices **briefly**.

name | section | date

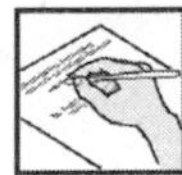

Pre-Lab Exercises for Experiment 3

These exercises are to be completed after you have read the experiment but before you come to the laboratory to perform it.

1. Pat Student weighed a 25-mL graduated cylinder, filled it to the 10-mL mark with dry sand and weighed the sand-filled cylinder. Next, Pat used a pipet to add 10.00 mL of water to the sand. Pat recorded the volume of the sand and water mixture after it had been stirred to allow the water to penetrate to the bottom of the sand column. Use Pat's data to find (a) the apparent density of the dry sand and (b) the actual density of the sand particles.

Mass of cylinder plus sand	37.5783 g
Mass of empty cylinder	22.1652 g
Volume of sand and water	17.50 mL

2. The radius of a magnesium atom is 1.60×10^{-8} cm. What is the volume of 1.00 mol of magnesium atoms?

name ______ section ______ date ______

Summary Report on Experiment 3

Mass of aluminum cylinder ______

Height of aluminum cylinder ______

Radius of aluminum cylinder ______

Final buret reading ______

Initial buret reading ______

Volume of water delivered ______

Calibration mark on graduated cylinder touched by meniscus ______

Mass of graduated cylinder plus sand ______

Mass of empty graduated cylinder ______

Mass of sand ______

Final buret reading ______

Initial buret reading ______

Volume of water delivered ______

Calibration mark on graduated cylinder touched by meniscus ______

Mass of graduated cylinder plus table salt ______

Mass of empty graduated cylinder ______

Mass of table salt ______

Final buret reading ______

Initial buret reading ______

Volume of water delivered ______

Calibration mark on graduated cylinder touched by meniscus ______

Mass of graduated cylinder plus table salt ______

Mass of empty graduated cylinder ______

Mass of table salt ______

Volume of alcohol delivered from pipet ______

Calibration mark on graduated cylinder touched by meniscus ______

Results

Mass of aluminum cylinder ____________________

Volume of aluminum (based on dimensions) ____________________

Volume of aluminum (from water displacement) ____________________

Density of aluminum (*V* based on dimensions) ____________________

Density of aluminum (*V* from water displacement) ____________________

Mass of sand ____________________

Apparent volume of dry sand ____________________

Actual volume occupied by sand particles ____________________

Density of sand (based on apparent volume) ____________________

Density of sand (based on actual volume) ____________________

Mass of table salt ____________________

Apparent volume of dry table salt ____________________

Actual volume occupied by salt particles ____________________

Density of salt (based on apparent volume) ____________________

Density of salt (based on actual volume) ____________________

Mass of table salt recovered from evaporating dish ____________________

VSEPR

EXP
4

Laboratory Time Required

Three hours.

Special Equipment and Supplies

Balloons, 10″-diameter, helium-quality
Balloons, 13″-diameter, helium-quality
String
Scissors
Grommet
Tape measure

Objective

In performing this experiment, the student will gain insight into the use of the VSEPR model for predicting the shapes of covalent molecules and molecular ions.

Safety

This experiment does not expose students to chemical hazards. However, some institutions require that safety glasses or goggles must be worn whenever work is being done in a laboratory environment.

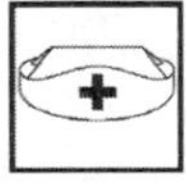

First Aid

It is not expected that any injuries could result from the performance of this experiment.

Preamble

It is often difficult for students to relate two-dimensional representations of molecules to the three-dimensional shapes of the molecules themselves. Yet an understanding of molecular shape is necessary for the prediction of many properties of molecules, such as polarity and Lewis basicity. In performing this experiment, the students will build models of molecules and probe the predictions of the VSEPR theory of molecular shape.

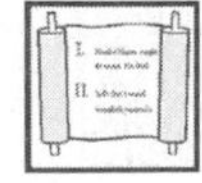

Principles

The first modern atomic theory, proposed by John Dalton in 1800, assumed that the properties of a molecule would be determined by the numbers and types of atoms that were combined within it. Dalton's theory marked the beginning of modern chemistry. Further refinements were made as subatomic particles were discovered and it was learned that atoms combine by the transfer of electrons from metal to non-metal in ionic compounds, or by the sharing of electrons between non-metals, in covalent compounds and polyatomic ions (also called molecular ions).

The properties of covalent molecules frequently depend not only on the number and types of atoms within the molecules but also on the shapes of the molecules. For instance, the linear arrangement of the atoms in a carbon dioxide molecule produces a non-polar molecule, which is a gas at room temperature and atmospheric pressure. Conversely, the bent arrangement of the atoms in a water molecule means that water is polar, giving it a comparatively high boiling point and making it an excellent solvent for most ionic compounds. The question is, "Why is CO_2 linear while H_2O is bent?"

One answer to that question is provided by VSEPR, the valence shell electron-pair repulsion theory. This theory starts with the idea that the central atom in a molecule is surrounded by bonding pairs (electrons shared by two atoms) and lone pairs (electrons attracted to the nucleus of a single atom) and that these pairs, all being negatively charged, will try to maintain the maximum possible distance between themselves. While it is not possible to locate electrons, experiments can locate the atoms in the molecule. The geometry of the molecule, defined by the relative positions of the atoms, will reflect the placement of the various electron pairs around the central atom.

In this experiment, small balloons, representing atoms bound to a central atom, will be mounted in a grommet, representing the central atom. "Inter-balloon distances" will be measured to determine the optimal geometry for the placement of 2, 3, 4, 5, and 6 atoms around a central atom. In addition, larger balloons, representing lone pairs, will then replace some of the smaller balloons. Again, measurement of "inter-balloon distances" will permit the determination of the optimum geometry for molecules in which lone pairs are present.

Procedure

Choose six balloons, each a different color. Blow the balloons up and tie each with a string of the same color as the balloon.

Draw the strings of two of your balloons through the grommet. Make the strings as tight as possible. Confirm that the balloons' tops are furthest apart when the two balloons lie along a straight line (180° angle).

Draw the string of a third balloon through the grommet. Make the strings as tight as possible. Confirm that the tops of the three balloons are the furthest apart when the balloons are arranged in a trigonal (also called "planar triangular") arrangement. The angle between any two of the balloons should be 120°.

Add a fourth balloon to the three in the grommet. Force the balloons into a square planar configuration (90° angles) and pull the strings tight. Measure the distance between the top of one balloon and the tops of each of the two balloons adjacent to it. Also, measure the distance between the top of the first balloon and the top of the balloon opposite it.

Now, adjust the four balloons to make a tetrahedral configuration (109.5 ° angles). Tighten the strings. Measure the distance between the top of one balloon and the tops of each of the other three. Which configuration (square planar or tetrahedral) keeps the balloons furthest apart?

Add a fifth balloon to the four in the grommet. Force the balloons into a square-based pyramid (90° angles). Pull the strings tight. Measure the distance between the top of the axial balloon and the tops of each of the four equatorial balloons adjacent to it. Then, measure the distance between the top of an equatorial balloon and the top of the equatorial balloon opposite from it. Finally, measure the distance between the first equatorial balloon and the two equatorial balloons adjacent to it.

Now, adjust the five balloons to make a trigonal bipyramid (equatorial angles of 120° and an axial angle of 180°). Tighten the strings. Measure the distance between the top of one axial balloon and the tops of each of the three equatorial balloons. Also, measure the distance between one axial balloon and the other. Finally, measure the distance between the top of one equatorial balloon and the tops of each of the other two. Which configuration (square-based pyramid or trigonal bipyramid) keeps the balloons furthest apart?

Add a sixth balloon to the five in the grommet. Force the balloons into an octahedral configuration (90° angles). Pull the strings tight. Measure the distance between the top of the axial balloon and the tops of each of the four equatorial balloons adjacent to it. Then, measure the distance between the top of an equatorial balloon and the top of the equatorial balloon opposite from it. Also, measure the distance between the first equatorial balloon and the two equatorial balloons adjacent to it. Finally, measure the distance between the top of one axial balloon and the top of the other axial balloon.

Choose three larger balloons, each a different color. Blow the balloons up and tie each with a string of the same color as the balloon.

Replace one of the six balloons in the grommet with a larger balloon, which represents a lone pair. Tighten the strings and confirm that the five smaller balloons are arranged in a square-based pyramid.

Replace another of the six original balloons in the octahedron with another larger balloon. Tighten the strings. The larger balloons will be either adjacent to or opposite to each other. Confirm that the configuration in which the "lone pairs" make a 180° angle is the preferred configuration.

Remove one of the smaller balloons from the grommet. Tighten the strings. The positions of the lone pairs may vary. The possibilities are that they will both be equatorial, that they will both be axial, or that one will be axial and one will be equatorial. Force the balloons into each of these configurations. Measure the distance between the top of one large balloon and each of the other balloons in each of the possible configurations. State which is the preferred configuration.

Replace another of the smaller balloons in the grommet with the third large balloon. (There should now be three large balloons and two small ones in the grommet.) Tighten the strings. The large balloons again can be placed in a variety of positions (two axial and one equatorial; three equatorial; two equatorial and one axial). Force the balloons into each of these configurations. Measure the distance between the top of one large balloon and each of the other balloons in each of the possible configurations. State which is the preferred configuration.

Disposal of Reagents

The balloons and strings may be saved for use by subsequent classes or they may be deflated and disposed of in the containers for solid waste.

name section date

Pre-Lab Exercises for Experiment 4

These exercises are to be completed after you have read the experiment but before you come to the laboratory to perform it.

1. Consult your textbook or other introductory chemistry texts and supply formulas for the molecules or molecular ions described below:

   a. a linear, non-polar molecule containing carbon and sulfur atoms

   b. a tetrahedral, molecular ion containing nitrogen and hydrogen atoms

   c. a square-planar molecule containing xenon and fluorine atoms

2. Draw the Lewis structures and predict the geometries of the molecules and molecular ions identified below.

   a. PCl_5

   b. ClO_4^-

   c. SF_4

3. Sketch a molecule of SBr_4Cl_2 in which the chlorines are in axial positions. Sketch another molecule of SBr_4Cl_2 in which the chlorines are in equatorial positions. What is the Cl–S–Cl angle when the chlorines are in the axial position? In the equatorial position?

4. Sketch a molecule of PBr_3Cl_2 in which the chlorines are in axial positions. Sketch another molecule of PBr_3Cl_2 in which the chlorines are in equatorial positions. What is the Cl–P–Cl angle when the chlorines are in the axial position? In the equatorial position?

name ______ section ______ date ______

Summary Report on Experiment 4

Use the spaces given below to record your data on the "inter-balloon distances."

A. *Configurations with small balloons only*

1. Two balloons

Color of first balloon ______ Color of second balloon ______

Inter-balloon distance ______

2. Three balloons

Color of first balloon ______

Color of second balloon ______ Color of third balloon ______

Inter-balloon distances (______ to ______) ______

(______ to ______) ______

(______ to ______) ______

3. Four balloons in square-planar arrangement

Color of first balloon ______

Balloons adjacent to first balloon

Color of second balloon ______ Color of third balloon ______

Balloon opposite first balloon

Color of fourth balloon ______

Inter-balloon distances (______ to ______) ______

(______ to ______) ______

(______ to ______) ______

(______ to ______) ______

(______ to ______) ______

(______ to ______) ______

4. Four balloons in tetrahedral arrangement

Color of first balloon ____________ Color of second balloon ____________

Color of third balloon ____________ Color of fourth balloon ____________

Inter-balloon distances (____________ to ____________) ____________

(____________ to ____________) ____________

(____________ to ____________) ____________

(____________ to ____________) ____________

(____________ to ____________) ____________

(____________ to ____________) ____________

(____________ to ____________) ____________

(____________ to ____________) ____________

5. Five balloons in square-planar arrangement

Color of first (axial) balloon ____________

Equatorial balloons

Color of first balloon ____________ Color of second balloon ____________

Color of third balloon ____________ Color of fourth balloon ____________

Inter-balloon distances (____________ to ____________) ____________

(____________ to ____________) ____________

(____________ to ____________) ____________

(____________ to ____________) ____________

(____________ to ____________) ____________

(____________ to ____________) ____________

(____________ to ____________) ____________

(____________ to ____________) ____________

(____________ to ____________) ____________

(____________ to ____________) ____________

name ______ section ______ date ______

6. Five balloons in trigonal-bipryramidal arrangement

Axial balloons

Color of first balloon ______ Color of second balloon ______

Equatorial balloons

Color of first balloon ______ Color of second balloon ______

Color of third balloon ______

Inter-balloon distances (______ to ______) ______

(______ to ______) ______

(______ to ______) ______

(______ to ______) ______

(______ to ______) ______

(______ to ______) ______

(______ to ______) ______

(______ to ______) ______

(______ to ______) ______

7. Six balloons

Axial balloons

Color of first balloon ____________________ Color of second balloon ____________________

Equatorial balloons

Color of first balloon ____________________ Color of second balloon ____________________

Color of third balloon ____________________ Color of fourth balloon ____________________

Inter-balloon distances (____________ to ____________) ____________

(____________ to ____________) ____________

(____________ to ____________) ____________

(____________ to ____________) ____________

(____________ to ____________) ____________

(____________ to ____________) ____________

(____________ to ____________) ____________

(____________ to ____________) ____________

(____________ to ____________) ____________

name ______ section ______ date ______

B. *Configurations with large and small balloons*

1. Five small balloons and one large balloon

Axial balloons

Color of large balloon ______ Color of small balloon ______

Equatorial balloons

Color of first balloon ______ Color of second balloon ______

Color of third balloon ______ Color of fourth balloon ______

Inter-balloon distances (______ to ______) ______

(______ to ______) ______

(______ to ______) ______

(______ to ______) ______

(______ to ______) ______

(______ to ______) ______

(______ to ______) ______

(______ to ______) ______

(______ to ______) ______

2. Three small balloons and two large balloons

Large balloons axial

Color of first balloon ____________________ Color of second balloon ________________

Small balloons equatorial

Color of first balloon ____________________ Color of second balloon ________________

Color of third balloon ____________________

Inter-balloon distances (axial to axial) ____________________

(axial to equatorial) ____________________

(equatorial to equatorial) ____________________

Small balloons axial

Color of first balloon ____________________ Color of second balloon ________________

Large balloons equatorial

Color of first large balloon ________________

Color of second large balloon ______________

Color of small equatorial balloon ___________

Inter-balloon distances (axial to axial) ____________________

(axial to large equatorial) ____________________

(axial to small equatorial) ____________________

(large equatorial to small equatorial) ____________________

name ______________________ section ______________________ date __________

One large balloon axial and one large balloon equatorial

Axial

Color of large balloon ______________ Color of small balloon ______________

Equatorial

Color of large balloon ______________

Color of first small balloon ______________

Color of second small balloon ______________

Inter-balloon distances (axial to axial) ______________

(small axial to small equatorial) ______________

(large axial to small equatorial) ______________

(small axial to large equatorial) ______________

(large axial to large equatorial) ______________

(small equatorial to small equatorial) ______________

(large equatorial to small equatorial) ______________

(large equatorial to large equatorial) ______________

3. Two small balloons and three large balloons

Large balloons axial

Color of first balloon ____________________ Color of second balloon ____________________

Equatorial

Color of large balloon ____________________

Color of first small balloon ____________________

Color of second small balloon ____________________

Inter-balloon distances (axial to axial) ____________________

(axial to large equatorial) ____________________

(axial to small equatorial) ____________________

(small equatorial to small equatorial) ____________________

(small equatorial to large equatorial) ____________________

(large equatorial to large equatorial) ____________________

Small balloons axial

Color of first balloon ____________________ Color of second balloon ____________________

Large balloons equatorial

Color of first balloon ____________________ Color of second balloon ____________________

Color of third balloon ____________________

Inter-balloon distances (axial to axial) ____________________

(axial to equatorial) ____________________

(equatorial to equatorial) ____________________

name ______________________ section ______________________ date __________

One large balloon axial and one large balloon equatorial

Axial

Color of large balloon ______________________ Color of small balloon ______________________

Equatorial

Color of small balloon ______________________

Color of first large balloon __________________

Color of second large balloon ________________

Inter-balloon distances (axial to axial) ______________________

(small axial to small equatorial) ______________________

(large axial to small equatorial) ______________________

(small axial to large equatorial) ______________________

(large axial to large equatorial) ______________________

(small equatorial to small equatorial) ______________________

(large equatorial to small equatorial) ______________________

(large equatorial to large equatorial) ______________________

EXP 5

Separation of a Mixture into Its Components by Fractional Crystallization

Laboratory Time Required Three hours.

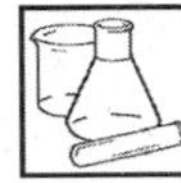

Special Equipment and Supplies

Hot plate
Buchner funnel and vacuum flask
Ice bath
3 M sulfuric acid, $H_2SO_4(aq)$
Salicylic acid/$CuSO_4 \cdot 5H_2O$ mixture
Boiling chips
95% ethanol, $C_2H_5OH(l)$

Objective In the course of this experiment, students will separate mixtures of salicylic acid and copper sulfate pentahydrate into their components.

Safety

The usual precautions should be taken to prevent exposure of the skin or eyes to chemicals.

Boiling solutions may bump suddenly, splashing hot liquid over a wide area. Add boiling chips to solutions before heating is begun to prevent overheating and the resulting bumping. Beware of hot liquids that are still; they may be superheated.

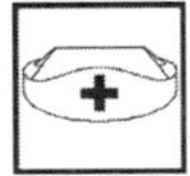

First Aid

Rinse chemicals from the skin with water.

For treatment of severe burns, see a doctor.

Preamble

The resolution of mixtures into their components is an important part of experiments concerned with analysis or synthesis. **Analysis** is concerned with finding the quantitative composition of a mixture. **Synthesis** is the formation of a substance from simpler substances. Synthesis usually includes a separation to isolate the desired product from large amounts of solvent and by-products. This separation is often followed by a purification step that removes trace amounts of other substances from the final product.

Some of the experimental techniques that may be used to separate or purify mixtures are: fractional crystallization, fractional distillation, chromatography, and zone refining. Each of these makes use of small differences in a physical property such as solubility, vapor pressure, melting point, or tendency to be adsorbed on a surface, to separate the components of a mixture. In this experiment, you will use fractional crystallization to resolve a mixture of salicylic acid and $CuSO_4 \cdot 5H_2O$ into its components.

Principles

The oft-quoted rule of thumb "like dissolves like" suggests that the ionic salt $CuSO_4 \cdot 5H_2O$ would be more soluble in the highly polar solvent H_2O than in a solvent of lower polarity, such as ethanol. For the organic solute salicylic acid (2-hydroxybenzoic acid), these solubility tendencies might well be reversed. Such is indeed the case.

Figure 5.1 shows the solubility (in g solute per 100 g of H_2O) of $CuSO_4 \cdot 5H_2O$ to be one to two orders of magnitude higher than that of salicylic acid. Further, the solubilities of both solutes are seen to increase with temperature. A mixture of these two solutes could be dissolved in 100 mL of water at 75°C to 100°C. Most of the salicylic acid could then be precipitated as fine, off-white needles when the solution is cooled to 0°C. Up to 25 or 30 g of $CuSO_4 \cdot 5H_2O$ could remain in the cooled solution without precipitating. This procedure will be used in this experiment to separate the salicylic acid in good yield and purity.

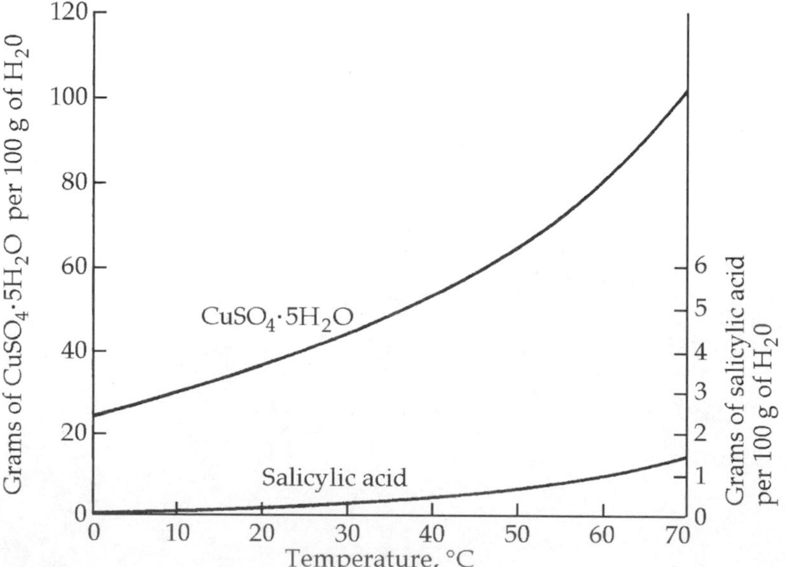

Figure 5.1 The solubility in water of $CuSO_4 \cdot 5H_2O$ (left scale) and salicylic acid (right scale)

It is harder to crystallize $CuSO_4 \cdot 5H_2O$ from water, especially if we wish to prevent the precipitation of the small amount of salicylic acid left in solution from the first crystallization. We can obtain pure $CuSO_4 \cdot 5H_2O$ by carefully evaporating the solution to about 25 mL and then adding an equal volume of 95% ethanol. This lowers the dielectric constant of the solvent medium so that essentially all of the $CuSO_4 \cdot 5H_2O$ precipitates as fine blue crystals when the solution is cooled to room temperature or lower. At the same time, the salicylic acid will remain in solution.

Although both precipitates may appear pure, an analysis would no doubt show that the salicylic acid is contaminated by trace amounts of $CuSO_4 \cdot 5H_2O$, and vice versa. If, for some reason, we need ultrapure salicylic acid or $CuSO_4 \cdot 5H_2O$, we obviously could repeat the dissolution and crystallization procedure to obtain recrystallized products. That will not be necessary for this experiment.

Procedure

Obtain an unknown mixture and record its number on the Summary Report Sheet. Add about 2.5 grams of the unknown mixture to a previously weighed, clean, and dry, 100-mL beaker. Weigh the beaker containing the mixture and calculate the mass of the unknown.

Add 50 mL of water containing about 1 mL of 3 M H_2SO_4 solution to the beaker. The acid will prevent any possible chemical reaction between the copper sulfate and the salicylic acid. Heat the mixture on a hot plate, constantly stirring until all of the solid material has dissolved. Avoid boiling the solution. Remove the beaker from the heat, cover it with a watch glass, and allow it to cool enough to be comfortable to the touch.

Caution!

WAIT A FEW MINUTES BEFORE TRYING TO TOUCH THE BEAKER.

Procedure in a Nutshell

Dissolve a mixture of salicylic acid and copper sulfate pentahydrate in acidified water at elevated temperature. Cool the mixture in an ice-water bath to bring the salicylic acid out of solution. Separate the salicylic acid crystals from the copper sulfate solution by vacuum filtration. Reduce the volume of the remaining solution via boiling. After the heating has been stopped, add ethanol to the mixture to precipitate the copper sulfate pentahydrate. Collect these crystals by filtration.

Place the beaker in an ice bath. A large quantity of salicylic acid needles should settle to the bottom of the beaker. Keep the beaker in the ice bath until crystallization is complete. Set up a vacuum filtration apparatus as shown in Figure 5.2 and place a piece of filter paper in the Buchner funnel. Swirl the beaker and pour its contents into the Buchner funnel. Transfer the salicylic acid crystals in this way as completely as possible to the filter paper. A minimum amount of cold rinse water may be used to complete the transfer. Pour the blue solution collected in the filtration flask into a second clean beaker and reserve it for the crystallization of $CuSO_4 \cdot 5H_2O$. Draw air through the Buchner funnel for several minutes to remove as much water as possible; then remove the filter paper and its mat of salicylic acid crystals. Carefully transfer the crystals to a dry, preweighed filter paper to finish drying. At the end of the lab period, weigh the paper and its crop of crystals and calculate the mass of salicylic acid recovered. Also report the percent by mass of salicylic acid in the original mixture.

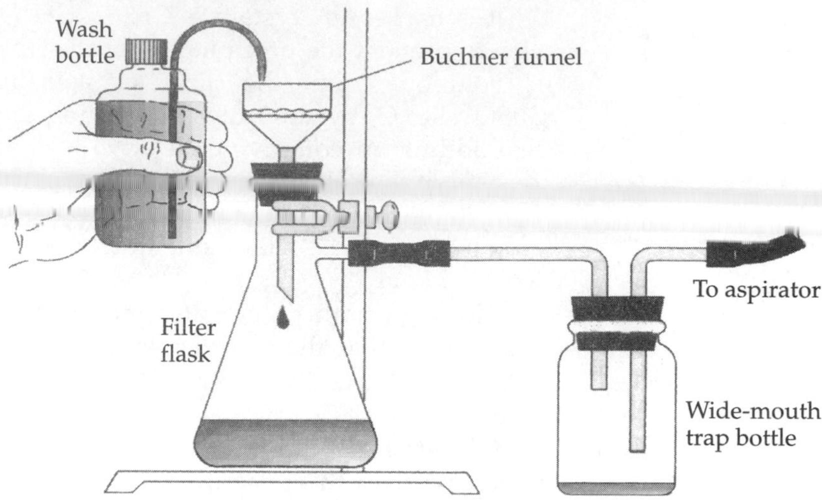

Figure 5.2 Vacuum filtration apparatus

Place the beaker containing the blue copper sulfate solution on a hot plate, add a stirring rod and one or two boiling chips to the beaker, and boil the solution gently until its volume has been reduced to about 25 mL.

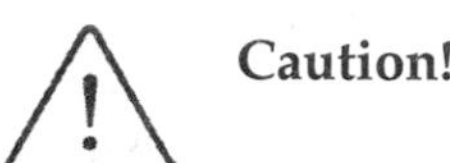

Caution!

THE STIRRING ROD AND THE BOILING CHIPS ARE INTENDED TO BE A SOURCE OF BUBBLES THAT WILL PREVENT SUPERHEATING, BUT BE VERY CAUTIOUS OF THE HOT SOLUTION ANYWAY, ESPECIALLY IF NO BUBBLES ARE OBSERVED. IF THE SOLUTION DOES BECOME SUPERHEATED, ANY DISTURBANCE COULD CAUSE IT TO BOIL OVER VIOLENTLY, RISKING INJURY TO PERSONS NEARBY.

Remove the beaker from the hot plate and, while stirring, slowly add about 15 mL of ethanol, or a sufficient volume to just cause the solution to become permanently cloudy. Cool the beaker to room temperature, then further cool it in an ice bath. A nice crop of blue $CuSO_4 \cdot 5H_2O$ crystals should settle to the bottom of the beaker. Collect the copper sulfate crystals on a preweighed filter paper as before, using the Buchner funnel. Rinse the product with a small amount of ethanol and draw air through it to dry it, using the vacuum filtration apparatus. Weigh the dry product and report the mass of $CuSO_4 \cdot 5H_2O$ recovered, as well as its percentage in the original mixture. (If the mass of the crystals seems to fall constantly, the $CuSO_4 \cdot 5H_2O$ may be contaminated with ethanol. Allow additional time for the ethanol to evaporate before attempting to weigh the crystals again.)

Comment briefly on the appearance of the products. Is there any visual evidence that they are not pure?

Disposal of Reagents

Place the $CuSO_4 \cdot 5H_2O$ and the salicylic acid, respectively, in the labeled collection bottles. All other chemicals, including the filtrate, may be rinsed down the drain.

name section date

Pre-Lab Exercises for Experiment 5

These exercises are to be completed after you have read the experiment but before you come to the laboratory to perform it.

1. Explain briefly why $CuSO_4$ is more soluble in water than in an organic (nonpolar) solvent. Why is the reverse true for salicylic acid?

2. Suggest reasons why the solubility of most solid substances in liquid solvents increases with temperature, whereas the solubility of gases decreases with temperature.

3. E.D. Student used 2.498 g of a mixture of copper sulfate pentahydrate and salicylic acid in the performance of this experiment. E.D. recovered 1.184 g of salicylic acid and 1.367 g of copper sulfate pentahydrate. What is wrong with E.D.'s results? How might this error have been avoided?

name section date

Summary Report on Experiment 5

Unknown number __________

Mass of unknown + beaker __________

Mass of beaker __________

Mass of unknown mixture __________

Salicylic Acid Crystallization

Mass of weighing paper + salicylic acid __________

Mass of weighing paper __________

Mass of salicylic acid __________

Calculation of percent salicylic acid in original mixture:

$CuSO_4 \cdot 5H_2O$ *Crystallization*

Mass of weighing paper + $CuSO_4 \cdot 5H_2O$ __________

Mass of weighing paper __________

Mass of $CuSO_4 \cdot 5H_2O$ __________

Calculation of percent $CuSO_4 \cdot 5H_2O$ in original mixture:

Percent of original sample not recovered __________

Comments concerning appearance and purity:

Fun with Solutions

EXP
6

Laboratory Time Required

Three hours.

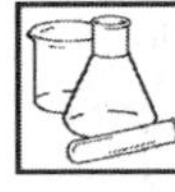

Special Equipment and Supplies

Paper plates
Cotton swabs
Plastic sandwich bags
Conductivity tester
Spray bottles
Labels
Dropper bottles
Phenolphthalein(*aq*)
Sodium thiosulfate, $Na_2S_2O_3\,(aq)$
Copper(II) sulfate, $CuSO_4\,(aq)$
Sodium hydroxide, $NaOH(aq)$
Iodine, $I_2\,(aq)$
Magnesium sulfate (anhydrous), $MgSO_4\,(s)$
Magnesium sulfate heptahydrate, $MgSO_4 \cdot 7H_2O\,(s)$
1 M barium chloride, $BaCl_2$
6 M ammonia, NH_3
0.1 M lead(II) nitrate, $Pb(NO_3)_2$
0.05 M cobalt(II) chloride, $CoCl_2$
0.05 M copper(II) sulfate, $CuSO_4$
0.05 M nickel(II) nitrate, $Ni(NO_3)_2$
Table salt, $NaCl(s)$
Table sugar, $C_{12}H_{22}O_{11}\,(s)$

Objective

In the course of performing this experiment, students will explore a number of physical and chemical changes. They will also learn to use the lab manual as a reference text and, ultimately, design simple chemical demonstrations.

Safety

Avoid getting acids or bases in your eyes or on your skin.

After completing the experiment, wash your hands with soap and water to remove any traces of toxic chemicals, such as the salts of lead, cobalt, or nickel.

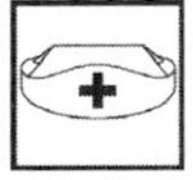

First Aid

Following skin contact with any of the reagents, wash the area with copious amounts of water.

If acid or base enters your eyes, use the eyewash fountain to flush away the chemical. Then, consult a physician.

Preamble

Two of the hardest skills for students to master in introductory chemistry classes are the ability to determine which chemical changes account for the observations they make when a reaction occurs, and the ability to represent those reactions with balanced chemical equations. This experiment allows students to explore the properties of solutions in an entertaining setting, provides support for distinguishing between physical changes and chemical changes, and introduces students to the processes of writing equations and designing experiments.

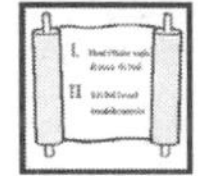

Principles

A magic show can be appreciated on several levels. A young child may believe that a magician has really pulled a rabbit out of an empty hat. An older child may sense that the magician probably has not really managed to change a scarf into a dove, but may not be able to offer a plausible explanation for the magician's illusion. Someone who has practiced sleight-of-hand may indeed know how the trick is done and still be able to enjoy seeing a talented practitioner perform the trick skillfully. In this experiment, you will have the opportunity to learn the chemistry behind some things that might appear in a "chemical magic show." You will prepare "secret" messages by writing in "invisible" ink. You will make mixtures that get hot or cold. You will learn how to distinguish sugar from salt without tasting either. You will find a liquid than can "pour out" a variety of colors, and mix colors that will disappear before your very eyes. Some of these "tricks" will involve physical changes alone; others will involve chemical reactions. Part of your job will be to determine which is which. In the course of finding the answers, you will learn to use this manual and your text as research tools and, ultimately, be able to design some "tricks" of your own.

"Invisible" inks can be prepared from a variety of substances. You will use three different kinds. One ink is an indicator, a substance that changes color as the acidity of the medium in which it is dissolved changes. You can read about indicators in Experiments 7, 8, 14, 19, 23, and 26 of this manual. You could also look into your textbook's chapters on acids and bases and on titrations or look up indicators in the text's index.

Another "invisible" ink works because of an oxidation/reduction reaction in which iodine (or its aqueous form, the triiodide ion) is reduced by the action of sodium thiosulfate. References to this reaction are found in Experiments 21 and 25. You might find further information by checking your text's chapter on redox reactions or looking for "thiosulfate" or "iodometric titrations" in the index.

The "magic" of the third "invisible" ink uses the reaction between copper ions and ammonia to form a complex ion. This phenomenon is mentioned in Experiments 11, 17, and 27. You could also find information by checking your text's sections on coordination compounds or complex ions. Complex ions are sometimes discussed in connection with the solubility of salts in water or in connection with transition metal chemistry.

The process of dissolving a salt (an ionic compound) in water is sometimes accompanied by the generation of a significant quantity of heat. Such a process is said to be "exothermic" and may be used to make chemical hot packs. Some salts dissolve in an "endothermic" manner, meaning that energy is absorbed from the surroundings as the salt dissolves. An endothermic dissolution process could be used to prepare a chemical cold pack. The generation of heat when a salt is dissolved is used for the identification of an unknown in Experiment 26. The amount of heat absorbed when another salt dissolves is measured in Experiment 23. For a more complete discussion, consult the sections of your textbook that deal with preparing solutions or with thermochemistry.

To illustrate that creating the hot or cold pack has not altered the salts, they are subjected to identical chemical reactions. The reaction of ammonia with the dissolved salts depends on the ability of ammonia to act as a base in aqueous solution. This reaction is used in Experiment 26. Ammonia's ability to act as a base is discussed in your textbook in the sections on acid/base chemistry and/or on the Bronsted/Lowry model. The reaction with barium chloride involves the formation of an insoluble barium compound. Experiments 24 and 27 involve precipitations of barium salts. You should be able to find further information in your text's chapter on solubility or in a section on precipitation or metathesis (double displacement) reactions.

It would be easy to build a machine that could tell sugar from salt. Table sugar is a nonelectrolyte, which has a large molar mass. Table salt is an electrolyte, which has a much lower molar mass than table sugar. Thus, one could distinguish sugar from salt by measuring the conductivity of an aqueous solution of an unknown crystal. Or, one could check the freezing point of solutions of equal masses of crystals, each dissolved in the same mass of water. The salt, because it dissociates and because it has a lower molar mass, would have the lower freezing point. The phenomenon of freezing point depression is discussed in Experiment 16 and the tests to distinguish sugar from salt are employed in Experiment 26. Your text has more information in the sections on electrolytic solutions, freezing point depression, or colligative properties.

Ammonia changes the colors of solutions by acting as a base (generating hydroxide ions that affect the colors of acid/base indicators or forming insoluble hydroxide compounds with metal ions) or by acting as a complexing agent (as it does with copper(II) ions). Sources of information on these phenomena have been cited above.

Many transition metal ions are highly colored. Mixing the appropriate amounts of solutions containing cobalt(II) ions, copper(II) ions, and nickel(II) can produce a startling result as the colors seem to disappear. The absorbance of light by colored solutions is discussed in Experiment 17. Your text may discuss color in the sections on the beginnings of quantum mechanics, on spectroscopy, or on transition metal chemistry.

Procedure

A. Invisible Inks

Using a pen or pencil, write "1," "2," and "3," respectively, on three paper plates. Also label each plate with your name. Dip a cotton swab into phenolphthalein solution. Write a secret message on plate #1 with the swab. Dip the swab in phenolphthalein as many times as is necessary to complete the writing. Set the plate aside to dry.

Repeat the procedure described above, using a new cotton swab, dipped in thiosulfate solution, to draw a picture on plate #2.

Repeat the procedure described above, using a new cotton swab, dipped in copper sulfate solution, to draw a picture on plate #3.

When the plates are dry, lightly spray the first with sodium hydroxide solution, the second with iodine solution, and the third with ammonia solution. Describe the appearance of each plate on the Summary Report Sheet.

Procedure in a Nutshell

Write invisible messages and use "developers" to make them visible. Show that the process of dissolution can generate or consume a considerable amount of energy and that solutes retain their distinctive chemical properties when in solution. Show that salt and sugar can be distinguished either by their conductive powers in solution and by their differing ability to depress the freezing point of water. Use a single reagent to make three different colors via reaction with three different substances. Combine colored solutions to give new shades (and create a "colorless" mixture).

B. *Hot and Cold*

Fill the bottom of a plastic sandwich bag with a thin layer of epsom salt (magnesium sulfate heptahydrate). Add 10 mL of water. Note whether the bag gets hot or cold.

Fill the bottom of a second sandwich bag with a thin layer of anhydrous magnesium sulfate. Add 10 mL of water. Note whether the bag gets hot or cold.

Place a pea-sized amount of epsom salt in each of two clean test tubes. Add 5 mL of water to each tube and stir the contents until the solid dissolves. Add one dropperful of 6 M ammonia to one tube and one dropperful of 1 M barium chloride to the other tube. Note your observations on the Summary Report Sheet.

Place a pea-sized amount of anhydrous magnesium sulfate in each of two clean test tubes. Add 5 mL of water to each tube and stir the contents until the solid dissolves. Add one dropperful of 6 M ammonia to one tube and one dropperful of 1 M barium chloride to the other tube. Note your observations on the Summary Report Sheet.

C. *Is It Sugar or Is It Salt?*

Weigh out 4 g of table sugar and place the sample in a small beaker. Add 10 mL of water and stir the mixture to dissolve the solid. Weigh out 4 g of table salt and place the sample in a second small beaker. Add 10 mL of water and stir the mixture to dissolve the solid.

Test the sugar and salt solutions you've just prepared for electrical conductivity. Obtain a conductivity tester and secure the battery connection. Place the copper leads in the sugar solution, and be sure that the leads do not touch each other. Note your observations on the Summary Report Sheet. Rinse the leads with distilled water and dry them. Repeat the conductivity test with the salt solution. Note your observations on the Summary Report Sheet. Clean the leads and disconnect the battery.

Next, determine whether the sugar solution or the salt solution has the lower freezing point. Prepare an ice/water/rock salt bath by placing equal amounts of each material in a large beaker. Stir the mixture well with a stirring rod or a thermometer. Be sure the temperature of the bath is –10°C or below. (If it is not, pour off some water and add more ice and rock salt; stir the mixture again before checking its temperature.) Place the beaker of sugar solution in the cold bath and stir the sugar solution with a thermometer until ice crystals begin to form or until the temperature of the solution reaches the temperature of the bath. Record your observations on the Summary Report Sheet. Rinse the thermometer and dry it. When the thermometer is again reading room temperature, repeat the procedure with the beaker containing the salt solution. Record your observations on the Summary Report Sheet.

D. Patriotic Liquid

Place a dropperful of phenolphthalein solution in a clean test tube. Add a dropperful of 6 M ammonia. Record your observations on the Summary Report Sheet.

Place a dropperful of lead(II) nitrate solution in a clean test tube. Add a dropperful of 6 M ammonia. Record your observations on the Summary Report Sheet.

Place a dropperful of copper(II) sulfate solution in a clean test tube. Add a dropperful of 6 M ammonia. Record your observations on the Summary Report Sheet.

E. Now You See It, Now You Don't

Place one dropperful of 0.05 M cobalt(II) chloride solution in a clean test tube. Note its color. Add 0.05 M copper(II) sulfate solution, dropwise, until the volume of the solution in the test tube has doubled. Note the color of the mixture. Add 0.05 M nickel(II) nitrate solution to the mixture, one drop at a time. Stir the mixture after each addition of nickel(II) nitrate solution. Add nickel(II) nitrate solution until the mixture appears colorless. On the Summary Report Sheet, note whether you have succeeded in producing a colorless mixture.

F. On Your Own

Design a chemical "magic trick" to demonstrate for your classmates. You might demonstrate how to make a blue message materialize on a white background, or how to have a white message appear on a brown background. You might prepare a cold pack or a hot pack using salts you've read about while doing research for this experiment. You might produce your school colors using a single liquid and various reagents, or you might devise a way for distinguishing between cornstarch and baking soda. Whether you devise a method for performing one of the suggested transformations, or devise something totally on your own, have your instructor check your idea for safety before you proceed.

Disposal of Reagents

Place any mixtures containing nickel, cobalt, or lead ions in the collection bottles provided. All other solutions and suspensions may be diluted and flushed down the drain with a good supply of water.

Questions

1. Explain each of the observations you've noted in performing this experiment with a sentence or two and/or an equation. Note which procedures involved chemical changes and which did not.

2. Explain your reasons for classifying certain changes as physical changes and others as chemical changes.

name _______________ section _______________ date _______________

Pre-Lab Exercises for Experiment 6

These exercises are to be completed after you have read the experiment but before you come to the laboratory to perform it.

1. Find the colors of the acid and base forms of phenolphthalein and two other indicators.

2. Complete and balance the equations started below.

a. ____ I_2 (*aq*) + ____ $S_2O_3^{2-}$ (*aq*) $\rightarrow$

b. ____ Cu^{2+} (*aq*) + ____ NH_3 (*aq*) $\rightarrow$

c. ____ Mg^{2+} (*aq*) + ____ OH^- (*aq*) $\rightarrow$

d. ____ Ba^{2+} (*aq*) + ____ SO_4^{2-} (*aq*) $\rightarrow$

3. Dilute solutions have physical properties that are close to those of the solvent. However, a solution's freezing point does differ from that of the solvent. Is the freezing point of a solution higher or lower than that of the solvent? Which factors determine how close the freezing point of the solution is to the freezing point of the solvent?

4. What are the "primary" colors?

name ______ section ______ date ______

Summary Report on Experiment 6

A. *Invisible Inks*

Mixture	*Observations*
henolphthalein/NaOH	______
Thiosulfate/iodine	______
Copper(II) sulfate/ammonia	______

B. *Hot and Cold*

Epsom salt plus water was ☐ hot ☐ cold

Anhydrous $MgSO_4$ plus water was ☐ hot ☐ cold

Mixture	*Observations*
NH_3 /epsom salt solution	______
$BaCl_2$ /epsom salt solution	______
NH_3 /solution of $MgSO_4$	______
$BaCl_2$ /solution of $MgSO_4$	______

C. *Is It Sugar or Is It Salt?*

Complete the table.

	Did the bulb light up?	*Did the solution freeze?*
Sugar solution	______	______
Salt solution	______	______

D. *Patriotic Liquid*

Mixture	*Colors*
Phenolphthalein/ NH_3	______
Lead(II) nitrate/ NH_3	______
Copper(II) sulfate/ NH_3	______

E. *Now You See It, Now You Don't*

What is the color of:

Cobalt(II) chloride (*aq*) ____________________

Nickel(II) nitrate (*aq*) ____________________

Copper(II) sulfate (*aq*) ____________________

Co^{2+} / Cu^{2+} mixture ____________________

Co^{2+} / Cu^{2+} / Ni^{2+} mixture ____________________

F. *On Your Own*

Outline the demonstration you wish to perform.

Properties of Oxides, Hydroxides, and Oxo-Acids

EXP
7

Laboratory Time Required

Three hours.

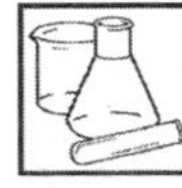

Special Equipment and Supplies

Red and blue litmus paper
Drinking straw
Glass tubes for gas generator
Sandpaper
Labels
Phenolphthalein indicator
Limewater, saturated $Ca(OH)_2(aq)$
Limestone chips, $CaCO_3(s)$
Sodium sulfite, $Na_2SO_3(s)$
Magnesium ribbon, Mg(*s*)
Granulated zinc, Zn(*s*)
Steel wool, Fe(*s*)
Light copper turnings, Cu(*s*)
Acetic acid, $HC_2H_3O_2(aq)$, 3 M
Sodium hydroxide, NaOH(*aq*), 3 M
Copper(II) sulfate, $CuSO_4(aq)$, 0.1 M
Aluminum sulfate, $Al_2(SO_4)_3(aq)$, 0.05 M
Iron(III) sulfate, $Fe_2(SO_4)_3(aq)$, 0.05 M
Ammonium chloride, $NH_4Cl(s)$

Objective

In performing this experiment, students will use various chemical tests to classify substances as acids or bases.

Safety

All of the acids, bases, and oxides should be regarded as **corrosive** or **caustic** and capable of causing burns to the skin or possible blindness.

Consequently, safety glasses with side shields and laboratory aprons are mandatory. Take every precaution against skin or eye exposure to chemicals.

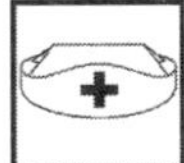

First Aid

Rinse the exposed skin or eyes thoroughly with water.

See a doctor if chemicals have entered the eye or if skin irritation is extensive.

Preamble

You will examine the physical and chemical properties of a number of compounds in this experiment. You will be asked to represent the observed reactions by balanced equations and to use the Periodic Law to account for trends and properties. A listing of equations for reactions you may encounter in performing this experiment is given in Table 7.1 at the end of the experiment.

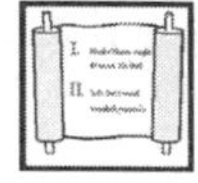

Principles

People have classified materials as acids or bases since ancient times, long before the development of the modern Atomic Theory. Acids characteristically have a sour taste. Limestone or marble will fizz when treated with acid substances. Bubbles form when acid is poured onto samples of zinc or iron. Many natural dyes take on characteristic colors in the presence of acids. For instance, purple grape juice turns red in acid solution. Red apple skins turn orange when treated with acids. Both the grape juice and the apple skins are acting as "indicators" in these cases.

Acids react readily with bases. Bases—or alkalis—characteristically have a bitter taste and a soapy feel. Addition of a base to ammonium salt will cause the release of ammonia gas. Like acids, bases cause indicators to take on distinctive coloration: grape juice turns green in basic solution; blueberry juice, which is dark red in the presence of acids, is blue in the presence of bases.

If base is added to an acid in the presence of an indicator, the color of the indicator will change. Red cabbage juice is orange-red in a strongly acidic environment. Its color will change to pinkish-red with the addition of small amounts of base. Continued addition of base will change the color to purple, green, yellow, and, finally, blue, in a strongly basic environment. These color changes are manifestations of the fact that an acid and a base react.

Once chemists accepted Dalton's Atomic Theory, it was natural that they attempt to discover the relationship between a substance's chemical formula and its acid or base properties. Many acids (today called "oxo-acids") contain oxygen and, in the late eighteenth century, it was thought that all acids were oxygen compounds; in fact, the German word for *oxygen* is *Sauerstoff*, meaning "acid material." The notion that all acids contain oxygen was proven wrong by Sir Humphrey Davy in 1810 when he demonstrated that "muriatic acid" (now known as hydrochloric acid) did not contain oxygen.

The first comprehensive acid/base theory in modern times was developed by Arrhenius in the late nineteenth century. The theory applies only to aqueous solutions, but is quite useful nonetheless, since water is an excellent solvent and many reactions take place in aqueous solution. According to the Arrhenius theory, in aqueous solution, an acid acts as a donor of protons (H^+ ions) and a base acts as a donor of hydroxide ions (OH^- ions). The reaction of an acid and a base is called "neutralization" and results in the formation of water and a salt. The reaction of nitric acid with sodium hydroxide is shown below in Equation 7.1; the products of this reaction are water and sodium nitrate.

$$HNO_3\ (aq) + NaOH\ (aq) \rightarrow H_2O\ (l) + NaNO_3\ (aq) \qquad (7.1)$$

Oxo-acids and hydroxides are ternary compounds composed of hydrogen, oxygen, and some other element. A number of such compounds can be obtained by the reaction of soluble, binary oxides with water, as shown in Equations 7.2 and 7.3. (Soluble, binary oxides are compounds of oxygen and another element that dissolve in water.)

$$N_2O_5\ (g) + H_2O\ (l) \rightarrow 2(NO_2)OH\ (aq) \tag{7.2}$$

$$Na_2O\ (s) + H_2O\ (l) \rightarrow 2NaOH\ (aq) \tag{7.3}$$

Examination of Equation 7.3 reveals that the reaction of sodium oxide with water has produced sodium hydroxide. Because sodium hydroxide is a base, sodium oxide is said to be a basic anhydride.

You may not readily recognize the product of the reaction shown in Equation 7.2, which shows that N_2O_5 is the acid anhydride of nitric acid (HNO_3). In Equation 7.2, the formula for nitric acid is written in an unusual way. The formula, $(NO_2)OH$, is a condensed structural formula, which tells us more than the molecular formula, HNO_3. All that the molecular formula reveals is that one hydrogen atom, one nitrogen atom, and three oxygen atoms somehow combine to make one molecule of nitric acid. The condensed structural formula tells us that, in nitric acid, a nitrogen atom is bonded directly to two oxygen atoms and to one hydroxyl ($-OH$) group. The hydroxide ion is basic, but the hydroxyl group is acidic.

All oxo-acids are in fact hydroxyl compounds. Writing condensed structural formulas for acids, such as $HC_2H_3O_2$ (acetic acid), can help us to understand why all hydrogens in oxo-acids are not necessarily acidic. The condensed structural formula for acetic acid is CH_3COOH. The central carbon in this molecule is bonded to a $-CH_3$ group, an oxygen atom, and a hydroxyl group. The hydrogen in the hydroxyl group is the acidic hydrogen; it may separate from the molecule, as a proton, in aqueous solution. The three hydrogens of the $-CH_3$ group do not separate from the carbon.

According to the Periodic Law, the properties of the elements vary in a periodic manner. These properties include the formulas of binary oxides, the tendency for atoms to acquire positive or negative charges when combined in molecules, and the tendency for binary oxides to act as anhydrides for acids, bases, or amphoteric substances. Amphoteric substances are substances that may act as either acids or bases. Because it is a characteristic of acids to react with bases, and vice versa, one criterion for calling a substance "amphoteric" is that the material reacts with both acids and bases. Thus, water-insoluble zinc hydroxide is amphoteric because it "dissolves" in both acidic and basic solutions. This dissolution process actually involves reaction with either the acid or the base, as is shown in Equations 7.4 and 7.5.

$$Zn(OH)_2\ (s) + 2HCl\ (aq) \rightarrow ZnCl_2\ (aq) + 2H_2O\ (l) \tag{7.4}$$

$$Zn(OH)_2\ (s) + 2NaOH\ (aq) \rightarrow Na_2Zn(OH)_4\ (aq) \tag{7.5}$$

The tendency for insoluble binary oxides to act as acid or base anhydrides is, likewise, determined by the ability of the insoluble oxide to dissolve in an acidic or basic solution. Silicon dioxide is shown to be an acid anhydride in Equations 7.6 and 7.7, whereas magnesium oxide is shown to be a basic anhydride in Equations 7.8 and 7.9. In Equations 7.6 and 7.9, NR signifies no reaction.

$$SiO_2\ (s) + HCl\ (aq) \rightarrow NR \qquad (7.6)$$

$$SiO_2\ (s) + 2NaOH(aq) \rightarrow Na_2SiO_3\ (aq) + H_2O\ (l) \qquad (7.7)$$

$$MgO\ (s) + 2HCl\ (aq) \rightarrow MgCl_2\ (aq) + H_2O\ (l) \qquad (7.8)$$

$$MgO\ (s) + NaOH\ (aq) \rightarrow NR \qquad (7.9)$$

Procedure

Procedure in a Nutshell

Find the acid and base colors of indicators and use an indicator to follow the neutralization reaction of an acid and base. Study the reaction of an acid with various metals. Determine whether limewater is acidic or basic. Generate carbon dioxide (CO_2) and determine whether it is an acidic or basic anhydride. Generate sulfur dioxide (SO_2) and determine whether it is an acidic or basic anhydride. Study the reactions of several salts with sodium hydroxide and determine whether the products of those reactions are acidic, basic, or amphoteric. Observe the effect of treating an ammonium salt with sodium hydroxide. Observe what happens when $MgSO_4$ and $MgSO_4 \cdot 7H_2O$ dissolve in water; also, note what happens when each of the solutions is treated with sodium hydroxide.

A. *Properties of Acids and Bases*

1. *Reactions with indicators.* Prepare about 3 mL each of 1 M acetic acid and 1 M sodium hydroxide in labeled test tubes by adding (with caution) 1 mL each of the 3 M stock solutions to 2 mL of water. Use a clean stirring rod to transfer a drop of the acetic acid solution to pieces of red and blue litmus paper. In a similar manner, place 1 drop of the sodium hydroxide solution on red and blue litmus papers. Report any color changes. Pour about 1 mL each of the acetic acid and sodium hydroxide solutions into separate test tubes and add 1 drop of phenolphthalein indicator solution to each test tube. Note the color of the indicator in the acidic and basic solutions, respectively.

2. *Neutralization.* Using an eyedropper or pipet, place 5 drops of the NaOH/phenolphthalein solution in a small beaker or test tube. Using a second eyedropper, add the CH_3COOH solution, drop by drop, with mixing, until the color just disappears. Note the color after each addition of acid. Write the equation for the reaction between acetic acid and sodium hydroxide.

3. *Reactions of acid with active metals.* Place four test tubes in a test tube rack. Label one tube "Mg" and place a small piece of magnesium ribbon in it. Label a second tube "Zn" and add one-half inch of granular zinc. Label a third tube "Cu" and add one-half inch of light copper turnings. Label a fourth tube "Fe" and put a small ball of steel wool in it. Add about 1 mL of 3 M acetic acid to each test tube. Note your observations. Write the chemical equations corresponding to the reactions observed.

B. *Properties of* CaO

Limewater is a solution that is prepared by dissolving 2 g of calcium oxide (called lime or quicklime) in a liter of water. Carefully pour 5 mL of the provided limewater into a 6-inch test tube without disturbing the solid at the bottom of the limewater container. If the solution obtained is not clear, filter it through filter paper as shown in Figure 7.1. Dip a clean stirring rod into the solution and touch the wet tip to pieces of red and blue litmus paper. Note your observations. Write an equation for the reaction of calcium oxide with water and state whether the product of the reaction is a (basic) hydroxide or an oxo-acid.

Put a drinking straw into the limewater that remains in the test tube. Gently blow through the straw into the limewater solution until you see a definite change in the appearance of the limewater. (You may need to keep exhaling into the limewater for a minute or two.) Record your observations. Set aside the test tube into which you have exhaled; it will be tested again in section C.

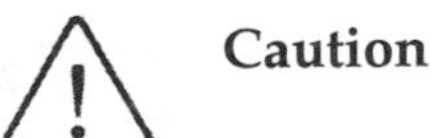

Caution! KEEP THE LIMEWATER OUT OF YOUR EYES!

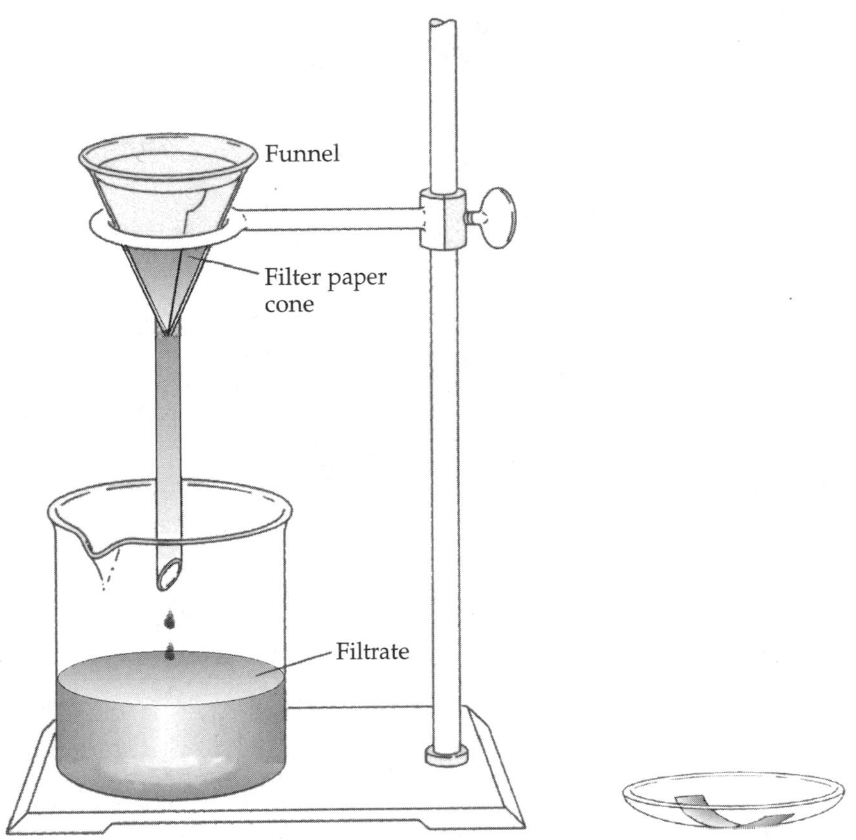

Figure 7.1 Gravity filtration apparatus

Figure 7.2 Litmus paper adhering to underside of watch glass

C. Properties of CO_2

Test the acidity of aqueous carbon dioxide as follows. Moisten a strip of blue litmus paper with water and drape it across the convex side of a watch glass as shown in Figure 7.2. Place about one spatulaful of granular $CaCO_3$ in a clean, dry test tube, and stand the test tube upright in a test tube rack. Carefully pour about 2 mL of 3 M acetic acid into the test tube without wetting the test tube lip. Cover the test tube with the watch glass so that the litmus paper is exposed to any gas generated in the reaction. Note any uniform color change of the litmus paper covering the test tube but ignore any random specks of color caused by acid splattering onto the litmus paper during the course of the reaction. Record your observations. Write an equation for the reaction of calcium carbonate with acetic acid and an equation accounting for the fizzing that occurs when carbonates are treated with acid. Also, write an equation for the reaction of carbon dioxide with water. Is the product of this reaction a (basic) hydroxide or an oxo-acid?

Rinse the test tube containing the calcium carbonate and acetic acid mixture. Pour off the rinse water, but retain the solid carbonate sample. Obtain a bent glass tube, a one-hole stopper or cork to fit your test tube, and a piece of rubber tubing about 1 foot long. These are to be connected as shown in Figure 7.3. Add about 2 mL of 3 M acetic acid to the test tube containing the $CaCO_3$, insert the stopper and exit tube, and allow any gas produced by the reaction of the calcium carbonate and the acid to bubble through the limewater for a few minutes. Record your observations.

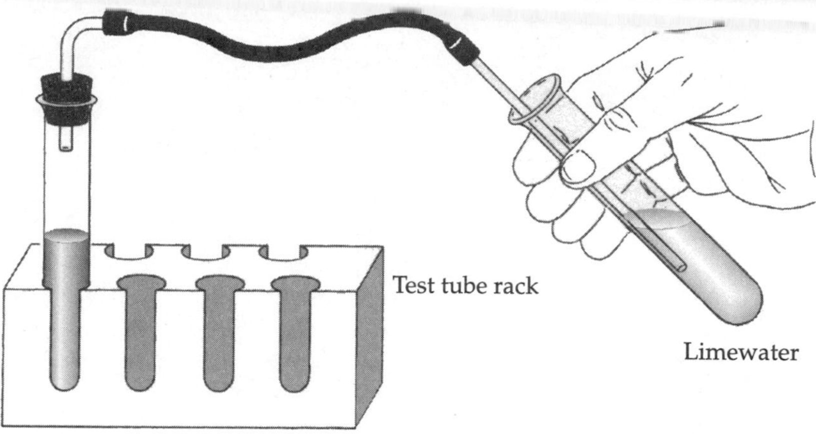

Figure 7.3 Limewater test for CO_2

Add about 5 mL of 3 M acetic acid to the test tubes containing the limewater through which the bubbles have passed and into which you have exhaled. Note any observation. Identify the product of the reaction between the gas and the limewater, and write an equation showing its formation.

D. Properties of SO_2

Place a spatulaful of sodium sulfite in a clean test tube and add about 2 mL of water. Stir to dissolve the solid, and use a stirring rod to place a drop of the solution on a piece of red litmus paper and on a piece of blue litmus paper. Record your observations. Stand the test tube upright in a test tube rack, and carefully pour about 3 mL of 3 M acetic acid into the test tube, taking care not to wet the test tube lip. Cover the test tube with a watch glass that has a fresh piece of moist blue litmus paper adhering to its convex side. Note any uniform color change of the litmus paper. Record your observations. Write an equation for the reaction of the sodium sulfite and the acetic acid, and an equation that explains the origin of the gas that arises in this reaction. Finally, write an equation showing the reaction between the gas and water. Is the product of that reaction a (basic) hydroxide or oxo-acid?

E. Reaction of Sodium Hydroxide with Aqueous Salt Solutions

Place six clean test tubes, each containing about 1 mL of water, in a test tube rack. Add 2 drops of 0.1 M $CuSO_4$ to each of the first two tubes; add 2 drops of 0.05 M $Al_2(SO_4)_3$ to each of the next two tubes; add 2 drops of 0.05 M $Fe_2(SO_4)_3$ to each of the last two tubes. Then, add 5 drops of 3 M NaOH to the first of the six test tubes, stirring the test tube and examining its contents for any sign of reaction after each addition of base. Record your observations and repeat the process of adding 5 drops of 3 M NaOH , stirring, and examining, until each of the six test tubes has been treated in this manner. (Be sure to clean your stirring rod before working with each new tube.)

Add 10 drops of 3 M acetic acid to the *first* tube of each of the three pairs of test tubes, so that you will be testing the product of the reaction of each salt and sodium hydroxide with acid. As you did before, stir the contents of the test tube and examine the contents of the tube after each addition of reagent. Note your observations.

Add 10 drops of 3 M sodium hydroxide to the *second* tube of each of the three pairs of test tubes, so that you will be testing the product of the reaction of each salt and sodium hydroxide with additional sodium hydroxide. As you did before, stir the contents of the test tube and examine the contents of the tube after each addition of reagent. Note your observations.

Write equations for the reaction of each of the salts with sodium hydroxide, and for the reaction of the product of these reactions with additional acid or base. State whether the product of the initial reaction was a (basic) hydroxide, an oxo-acid, or an amphoteric material.

F. *Reaction of Sodium Hydroxide with an Ammonium Salt*

Place a spatulaful of ammonium chloride in a clean test tube. Stand the test tube upright in a test tube rack and carefully pour about 3 mL of 3 M sodium hydroxide into the test tube, taking care not to wet the test tube lip. Cover the test tube with a watch glass that has a fresh piece of moist red litmus paper adhering to its convex side. Note any uniform color change of the litmus paper. Record your observations. Write an equation for the reaction of the sodium hydroxide and the ammonium chloride. Is the gaseous product basic or acidic?

G. *Reaction of Sodium Hydroxide with Magnesium Salts*

Place a spatulaful of anhydrous magnesium sulfate in a clean test tube. Stand the test tube upright in a test tube rack and carefully pour about 3 mL of deionized water into the test tube. Stir the mixture and touch the bottom of the test tube, carefully. Record your observations.

Place a spatulaful of epsom salt in another clean test tube. Stand the test tube upright in a test tube rack and carefully pour about 3 mL of deionized water into the test tube. Stir the mixture and touch the bottom of the test tube, carefully. Record your observations.

Add about 1 mL of 3 M sodium hydroxide to each of the test tubes containing aqueous magnesium sulfate. Note your observations. Add about 2 mL of 3 M acetic acid to each of these test tubes. Record your observations. Are the prodicts of the reaction between aqueous magnesium sulfate and sodium-hydroxide (basic) hydroxides or oxo-acids?

Disposal of Reagents

All solutions may be disposed of by being neutralized, diluted with water, and then flushed down the drain. Insoluble solids, such as limestone chips and strips of aluminum, copper, iron, and zinc should be rinsed, dried, and then placed in the labeled collection bottles for reuse.

Questions

1. The procedure of adding aqueous acid, drop by drop, to a solution of base, as was done in this experiment, is a rather crude titration. One purpose of a titration is to determine the molarity of an acid or base. Assume your CH_3COOH was 1 M. Find the molarity of the NaOH, based on your data for the neutralization portion of this experiment.

2. In this experiment, you worked directly with three anhydrides (CaO, CO_2, and SO_2). You have also been given information about the behavior of two other anhydrides, SiO_2 and MgO (see Equations 7.6 through 7.9). In addition, you performed tests on two hydroxides, $Al(OH)_3$ and $Cu(OH)_2$, for which the corresponding anhydrides are Al_2O_3 and CuO, respectively. Based on this information, state whether metal oxides are basic anhydrides or acid anhydrides. Also, state whether nonmetal oxides are basic anhydrides or acid anhydrides. Which oxide(s) dissolve(s) in water to form amphoteric substances? In which part of the Periodic Table are the elements whose oxides form amphoteric substances found?

Table 7.1 Equations for Selected Reactions

$2Al(OH)_3\ (s) + 6CH_3COOH\ (aq) \rightarrow 6H_2O\ (l) + 2Al(CH_3COO)_3\ (aq)$

$Al(OH)_3\ (s) + NaOH\ (aq) \rightarrow NaAl(OH)_4\ (aq)$

$Al_2(SO_4)_3\ (aq) + 6NaOH\ (aq) \rightarrow 2Al(OH)_3\ (s) + 3Na_2SO_4\ (aq)$

$CaO\ (s) + H_2O\ (l) \rightarrow Ca(OH)_2\ (aq)$

$Ca(OH)_2\ (aq) + CO_2\ (g) \rightarrow CaCO_3\ (s) + H_2O\ (l)$

$CaCO_3\ (s) + 2CH_3COOH\ (aq) \rightarrow Ca(CH_3COO)_2\ (aq) + CO(OH)_2\ (aq)$

$CH_3COOH\ (aq) + NaOH\ (aq) \rightarrow H_2O\ (l) + NaCH_3COO\ (aq)$

$CO(OH)_2\ (aq) \rightarrow H_2O\ (l) + CO_2\ (g)$

$CO_2\ (g) + H_2O\ (l) \rightarrow H^+\ (aq) + HCO_3^-\ (aq)$

$Cu\ (s) + CH_3COOH\ (aq) \rightarrow NR$

$Cu(OH)_2\ (s) + 2CH_3COOH\ (aq) \rightarrow 2H_2O\ (l) + Cu(CH_3COO)_2\ (aq)$

$Cu(OH)_2\ (s) + NaOH\ (aq) \rightarrow NR$

$CuSO_4\ (aq) + 2NaOH\ (aq) \rightarrow Cu(OH)_2\ (s) + Na_2SO_4\ (aq)$

$2Fe\ (s) + 6CH_3COOH\ (aq) \rightarrow 3H_2\ (g) + 2Fe(CH_3COO)_3\ (aq)$

$Fe(OH)(SO_4)\ (aq) + CH_3COOH\ (aq) \rightarrow H_2O\ (l) + Fe(CH_3COO)(SO_4)\ (aq)$

$Fe(OH)(SO_4)\ (aq) + NaOH\ (aq) \rightarrow NR$

$Fe(SO_4)_3\ (aq) + 6NaOH\ (aq) \rightarrow 2Fe(OH)(SO_4)\ (aq) + Na_2SO_4\ (aq)$

$Mg\ (s) + 2CH_3COOH\ (aq) \rightarrow H_2\ (aq) + Mg(CH_3COO)_2\ (aq)$

$MgSO_4\ (s) \rightarrow Mg^{2+}\ (aq) + SO_4^{2-}\ (aq) + \text{heat}$

$MgSO_4 \cdot 7H_2O\ (s) + \text{heat} \rightarrow Mg^{2+}\ (aq) + SO_4^{2-}\ (aq) + 7H_2O\ (l)$

$MgSO_4\ (aq) + 2NaOH\ (aq) \rightarrow Mg(OH)_2\ (s) + Na_2SO_4\ (aq) + \text{heat}$

$Mg(OH)_2\ (s) + 2CH_3COOH\ (aq) \rightarrow 2H_2O\ (l) + Mg(CH_3COO)_2\ (aq)$

$NH_4Cl\ (s) + NaOH\ (aq) \rightarrow NaCl\ (aq) + H_2O\ (l) + NH_3\ (g)$

$Na_2SO_3\ (s) + 2H_2O\ (l) \rightarrow SO(OH)_2\ (aq) + 2NaOH\ (aq)$

$Na_2SO_3\ (aq) + 2CH_3COOH\ (aq) \rightarrow SO(OH)_2\ (aq) + 2NaCH_3COO\ (aq)$

$SO(OH)_2\ (aq) \rightarrow H_2O\ (l) + SO_2\ (g)$

$SO_2\ (g) + H_2O\ (l) \rightarrow H^+\ (aq) + HSO_3^-\ (aq)$

$Zn\ (s) + 2CH_3COOH\ (aq) \rightarrow Zn(CH_3COO)_2\ (aq) + H_2\ (g)$

name section date

Pre-Lab Exercises for Experiment 7

These exercises are to be completed after you have read the experiment but before you come to the laboratory to perform it.

1. Define the terms "acid" and "base." Give examples of an oxo-acid and a (basic) hydroxide.

2. List two or more characteristic properties of acids. Do the same for bases.

3. Why does ammonia present a problem in the Arrhenius definitions of acids and bases?

name ______ section ______ date ______

Summary Report on Experiment 7

A. Properties of Acids and Bases

1. Indicator colors

	Red litmus	*Blue litmus*	*Phenolphthalein*
Acetic acid	______	______	______
Sodium hydroxide	______	______	______

2. Neutralization

Drops of acid added	*Color of NaOH/phenolphthalein*
______	______
______	______
______	______
______	______
______	______
______	______
______	______

3. Reactions of acid with active metals

Equation

Magnesium + acetic acid ______

Observations:

Copper + acetic acid ______

Observations:

Iron + acetic acid ______

Observations:

Zinc + acetic acid ______

Observations:

B. Properties of CaO

Effect of limewater on: red litmus paper ______________

blue litmus paper ______________

Equation for the reaction of CaO and water:

Is the product ☐ a hydroxide or ☐ an oxo-acid?

Effect of exhaling into limewater.

Effect of adding acid to aerated limewater.

C. Properties of CO_2

Reaction of calcium carbonate with acid

Observations:

Equations:

Effect of CO_2 on moist litmus paper:

Equation for the reaction of CO_2 with water:

Is the product ☐ a hydroxide or ☐ an oxo-acid?

Effect of bubbling CO_2 through limewater:

Effect of adding acid to limewater through which CO_2 has bubbled:

Equations:

name section date

D. *Properties of* SO_2

Dissolving sodium sulfite

Observations:

Equations:

Effect of SO_2 on moist litmus paper:

Equation for the reaction of SO_2 with water:

Is the product ☐ a hydroxide or ☐ an oxo-acid?

E. *Reaction of Sodium Hydroxide with Aqueous Salt Solutions*

Addition of NaOH to:

	Observations	*Equations*
$CuSO_4$		
$Al_2(SO_4)_3$		
$Fe_2(SO_4)_3$		

Addition of CH_3COOH to products of reaction of NaOH and:

	Observations	*Equations*
$CuSO_4$		
$Al_2(SO_4)_3$		
$Fe_2(SO_4)_3$		

Addition of NaOH to products of reaction of NaOH and:

	Observations	*Equations*
$CuSO_4$		
$Al_2(SO_4)_3$		
$Fe_2(SO_4)_3$		

F. *Reaction of Sodium Hydroxide with an Ammonium Salt*

Effect of product gas on moist litmus paper:

Equation for reaction of sodium hydroxide and ammonium chloride:

Is the product gas ☐ basic or ☐ acidic?

Equation:

G. *Reaction of Sodium Hydroxide with Magnesium Salts*

Effect of dissolving anhydrous magnesium sulfate in water:

Effect of dissolving epsom salt in water:

Equations for reactions:

Effects of adding sodium hydroxide to solutions:

Effects of adding acetic acid to solutions treated with sodium hydroxide:

Is the product that results from the addition of sodium hydroxide

☐ a hydroxide or ☐ an oxo–acid?

Equation:

EXP
8

Volumetric Analysis: Acid/Base Titration Using Indicators

Laboratory Time Required

Three hours.

Special Equipment and Supplies

- Analytical balance
- Buret
- Buret clamp
- 10-mL pipet
- Pipet bulb
- Thermometer
- Sodium hydroxide, 0.05 M NaOH
- Potassium hydrogen phthalate (KHP), solid
- Phenolphthalein indicator
- Vinegar unknowns

Objective

In performing this experiment, students will practice the important laboratory skills of titration, dilution, and pipetting.

Safety

Bases, such as sodium hydroxide, can cause skin burns and are especially hazardous to the eyes.

Although vinegar is a dilute solution of a weak acid, it is, nevertheless, advisable to avoid splashing it in the eyes.

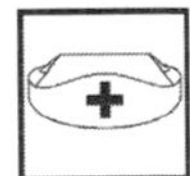

First Aid

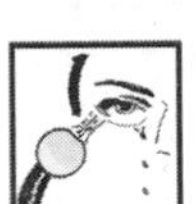

Following skin contact with sodium hydroxide, wash the area thoroughly with water. Should sodium hydroxide (or even vinegar) get in the eyes, rinse them thoroughly with water (at least 20 minutes of flushing with water is recommended) and seek medical attention.

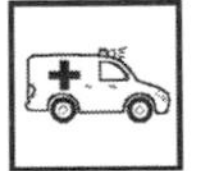

Preamble

In volumetric analysis, a known volume of a standard solution (one whose concentration is known) reacts with a known volume of a solution of unknown concentration. This procedure **standardizes** the latter solution, by allowing a calculation of its concentration.

The preparation and dispensing of solutions requires the use of calibrated glassware such as burets, pipets, and volumetric flasks. These items are illustrated in Figures I.3 through I.5, and instructions for using the equipment are provided in the Introduction. In this experiment, you will standardize a sodium hydroxide solution. You will then use this solution to analyze a solution containing an unknown concentration of acetic acid (the ingredient that gives vinegar its sour taste), using phenolphthalein as the indicator in the titration.

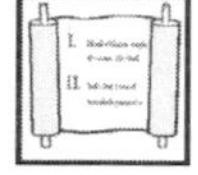

Principles

Titration

In a titration, a buret is used to dispense measured increments of one solution into a known volume of another solution. The object of the titration is the detection of the **equivalence point,** that point in the procedure where chemically equivalent amounts of the reactants have been mixed. Whether or not the equivalence point comes when equimolar amounts of reactants have been mixed depends on the stoichiometry of the reaction. In the reaction of acetic acid, $HC_2H_3O_2$, and NaOH, the equivalence point does occur when 1 mole of $HC_2H_3O_2$ has reacted with 1 mole of NaOH. However, in the reaction of H_2SO_4 and NaOH, the equivalence point occurs when 2 moles of NaOH have reacted with 1 mole of H_2SO_4.

The titration technique can be applied to many types of reactions, including oxidation-reduction, precipitation, complexation, and acid-base neutralization reactions.

Although a variety of instrumental methods for detecting equivalence points are now available, it is frequently more convenient to add an indicator to the reaction mixture. An indicator is a substance that undergoes a distinct color change at or near the equivalence point. The point in the titration at which the color change occurs is called the **end point**. Obviously, the titration will be accurate only if the end point and the equivalence point coincide fairly closely. For this reason, the indicator used in a titration must be selected carefully. Fortunately, a large number of indicators are commercially available and finding the right one for a particular titration is not a difficult task.

Acids and Bases

Although several definitions of acids and bases may be given, the classical Arrhenius concept will suffice for this experiment. According to this concept, an acid is a substance that dissociates in water to produce hydrogen ions; a base is a substance that dissociates in water to produce hydroxide ions. The classical Arrhenius acid-base reaction is one in which an acid reacts with a base to form water (from the combination of hydrogen ions and hydroxide ions) and a salt. Such a reaction is called a **neutralization reaction.** The neutralization reaction of acetic acid, the primary ingredient of vinegar, and sodium hydroxide is shown in Equation 8.1.

$$HC_2H_3O_2\ (aq) + NaOH\ (aq) \rightarrow H_2O\ (l) + NaC_2H_3O_2\ (aq) \qquad (8.1)$$

Because the mole ratio of acetic acid to sodium hydroxide is 1:1, the number of moles of acid present in the sample is equal to the number of moles of base that must be added to reach the equivalence point of the titration (see Equation 8.2).

$$M_b \times V_b = \text{moles base} = \text{moles acid} = M_a \times V_a \quad (8.2)$$

Indicators

Acid-base indicators are weak acids or bases that have different colors when in their dissociated and undissociated forms, respectively. The dissociation of the indicator HIn may be represented as shown in Equation 8.3.

$$\text{HIn} + \text{H}_2\text{O} \rightleftharpoons \text{H}_3\text{O}^+ + \text{In}^- \quad (8.3)$$

In acid solutions, the indicator exists predominantly as HIn. In basic solutions, it is present mainly as In^- ions, which impart a different color to the solution. Phenolphthalein is an indicator that is used frequently in acid-base titrations. It is colorless as long as the titration mixture contains an excess of acid and turns pink when the amount of base added slightly exceeds the amount of acid originally present. Thus, the end point will come shortly after the equivalence point. The conditions that cause phenolphthalein to change from its colorless (HIn) form to its colored (In^-) form are such that the difference between the end point and the equivalence point in a carefully performed titration is negligible.

Preparation of NaOH *Solution*

The general procedure for preparing carbonate-free NaOH solution involves decanting (pouring off) clear liquid from a 50% aqueous NaOH solution and then diluting this with freshly boiled and cooled water. This precaution is needed because atmospheric carbon dioxide tends to react with sodium hydroxide to form sodium carbonate. Because sodium carbonate has a very low solubility in 50% NaOH solution, any sodium carbonate present as an impurity in the solid NaOH used to prepare the solution (or formed by reaction of the solution of NaOH with carbon dioxide in the air) does not remain in solution. Rather, it slowly settles to the bottom of the solution as a white precipitate, from which the pure NaOH solution can be separated easily by decantation. The water used subsequently to dilute this solution should first be boiled to remove dissolved carbon dioxide.

Because of the very great hazard to the eyes presented by concentrated solutions of bases, you will not be given 50% NaOH. Rather, dilute (approximately 0.05 M), carbonate-free solution will be provided for your use. You will determine the exact concentration of NaOH in the approximately 0.05 M solution by titrating a known mass of potassium hydrogen phthalate. The reaction between these substances is shown in Equation 8.4.

$$\text{NaOH}\ (aq) + \text{KHC}_8\text{H}_4\text{O}_4\ (aq) \rightarrow \text{H}_2\text{O}\ (l) + \text{KNaC}_8\text{H}_4\text{O}_4\ (aq) \quad (8.4)$$

Thus, 1 mole of NaOH reacts exactly with 1 mole (204.22 g) of potassium hydrogen phthalate ($\text{KHC}_8\text{H}_4\text{O}_4$, generally labeled KHP). A variation on Equation 8.2 is used to determine the molarity of the sodium hydroxide solution from the data obtained in the standardization (see Equation 8.5)

$$M_b \times V_b = \text{moles base} = \text{moles acid} = (\text{mass of KHP}) \times \frac{1 \text{ mol KHP}}{204.22 \text{ g KHP}} \quad (8.5)$$

Procedure

Procedure in a Nutshell

Standardize a sodium hydroxide solution via titration of a known mass of KHP. Dilute an unknown volume of vinegar in a volumetric flask; mix well. Use a pipet to deliver 10.00 mL of the dilute vinegar to an Erlenmeyer flask. Titrate the diluted vinegar with the sodium hydroxide and determine its molarity.

Standardization of NaOH Solution

Clean a buret and prepare it for use in the titration according to the directions provided in the Introduction. Rinse the buret with small amounts of distilled water. Then rinse it twice with small portions of the NaOH solution provided. Finally, fill the buret with this solution.

Accurately weigh, to the nearest tenth of a milligram, 0.2 to 0.4 g of the dry KHP and add it to a 125-mL Erlenmeyer flask. Add 25 mL of distilled water and 2 to 3 drops of phenolphthalein indicator solution. Swirl to dissolve the solid. Read the initial volume of NaOH solution in the buret to the nearest hundredth of a milliliter. Titrate the KHP solution with the NaOH solution until a faint pink color, which does not disappear when the solution is mixed, is obtained. Read the final volume of NaOH solution in the buret to the nearest hundredth of a milliliter. Calculate the molarity of the NaOH solution. Repeat the standardization and average the results. If time permits, you may wish to perform a third standardization and use the criteria given in the Introduction to decide whether you should average the results of two or three trials.

Titration of an Unknown

Obtain an unknown volume of vinegar in a 100-mL volumetric flask. Dilute the solution to 100.00 mL with distilled water. Mix well. Rinse a 10-mL pipet with small portions of distilled water; then, use the pipet to deliver a 10-mL sample of water to a pre-weighed Erlenmeyer flask. Determine the mass and temperature of the water delivered; find the actual volume of water delivered by the pipet. Finally, rinse the pipet with small portions of your diluted vinegar solution, ultimately using the pipet to deliver a 10-mL sample of diluted vinegar to a 125-mL Erlenmeyer flask. Add 2 or 3 drops of phenolphthalein indicator and then titrate the acid with your 0.05 M NaOH solution until a faint, persistent pink color is obtained. Calculate the molarity of the acetic acid solution. Repeat the titration once or twice. Report the results of each titration and the average concentration of your acetic acid unknown.

Disposal of Reagents

Excess KHP should be placed in the containers used for solid waste. Solutions should be neutralized and diluted. They may then be flushed down the drain.

Questions

1. The letters TC and TD on volumetric glassware indicate whether it is calibrated to contain or to deliver a given volume. Which mode of calibration would be best for the following items as they are used in the experiment?

a. Buret

b. Volumetric flask

2. Graduated cylinders may be purchased marked either TC or TD. Briefly suggest a suitable use for each type.

3. A 50% NaOH solution has a density of 1.53 g/cm^3 and contains 50% NaOH by mass. What is the molar concentration of NaOH in this solution?

name section date

Pre-Lab Exercises for Experiment 8

These exercises are to be completed after you have read the experiment but before you come to the laboratory to perform it.

1. A 0.2816 g sample of KHP required 29.68 mL of sodium hydroxide to reach the phenolphthalein end point. Calculate the molarity of the NaOH solution.

2. A volume of 20.22 mL of the NaOH solution described in Question 1 was needed to titrate 10.06 mL of acetic acid to the phenolphthalein end point. Find the concentration of the acetic acid solution.

name ______ section ______ date ______

Summary Report on Experiment 8

Standardization of NaOH *Solution*

	Trial 1	*Trial 2*	*Trial 3**
Mass of KHP and container	______	______	______
Mass of container	______	______	______
Mass of KHP	______	______	______
Final buret reading, NaOH	______	______	______
Initial buret reading, NaOH	______	______	______
Volume used, NaOH	______	______	______
Molarity of NaOH solution	______	______	______
Average molarity of NaOH		______	

Titration of an Unknown

Unknown number	______
Mass of flask plus water	______
Mass of flask	______
Mass of water delivered by pipet	______
Temperature of water	______
Volume of diluted vinegar to be titrated	______

	Trial 1	*Trial 2*	*Trial 3**
Final buret reading, NaOH	______	______	______
Initial buret reading, NaOH	______	______	______
Volume used, NaOH	______	______	______
Molarity of acetic acid solution	______	______	______
Average molarity of acetic acid solution		______	

* Optional

EXP
8P

Volumetric Analysis: Acid-Base Titration[1]

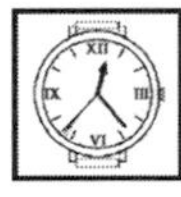

Laboratory Time Required Three hours.

Special Equipment and Supplies

CBL 2 interface
TI Graphing Calculator
DataMate Program
Analytical balance
Buret
Buret clamp
10-mL pipet
Pipet bulb
pH sensor
Temperature probe
Vinegar unknowns
Sodium hydroxide, 0.05 M NaOH
Potassium hydrogen phthalate (KHP), solid
Phenolphthalein indicator

Objective In performing this experiment, students will practice the important laboratory skills of titration, dilution, and pipetting.

Safety

Bases, such as sodium hydroxide, can cause skin burns and are especially hazardous to the eyes.

Although vinegar is a dilute solution of a weak acid, it is, nevertheless, advisable to avoid splashing it in the eyes.

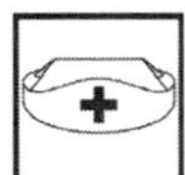

First Aid

Following skin contact with sodium hydroxide, wash the area thoroughly with water. Should sodium hydroxide (or even vinegar) get in the eyes, rinse them thoroughly with water (at least 20 minutes of flushing with water is recommended) and seek medical attention.

[1] Adapted from Holmquist, D.D., J. Randall, D.L. Votz, *Chemistry with Calculators: Chemistry Experiments Using Vernier Sensors with Texas Instruments CBL 2™* (Vernier Software and Technology, Beaverton, Oregon: 2000). Used with permission.

Preamble

In volumetric analysis, a known volume of a standard solution (one whose concentration is known) reacts with a known volume of a solution of unknown concentration. This procedure **standardizes** the latter solution, by allowing a calculation of its concentration.

The preparation and dispensing of solutions requires the use of calibrated glassware such as burets, pipets, and volumetric flasks. These items are illustrated in Figure I.3 , and instructions for using the equipment are provided in the Introduction. In this experiment, you will standardize a sodium hydroxide solution. You will then use this solution to analyze a solution containing an unknown concentration of acetic acid (the ingredient that gives vinegar its sour taste), using phenolphthalein as the indicator in the titration.

Principles

Titration

In a titration, a buret is used to dispense measured increments of one solution into a known volume of another solution. The object of the titration is the detection of the **equivalence point**, the point in the procedure where chemically equivalent amounts of the reactants have been mixed. Whether or not the equivalence point comes when equimolar amounts of reactants have been mixed depends on the stoichiometry of the reaction. In the reaction of acetic acid, $HC_2H_3O_2$ and NaOH, the equivalence point does occur when 1 mole of $HC_2H_3O_2$ has reacted with 1 mole of NaOH. However, in the reaction of H_2SO_4 and NaOH, the equivalence point occurs when 2 moles of NaOH have reacted with 1 mole of H_2SO_4

The titration technique can be applied to many types of reactions, including oxidation-reduction, precipitation, complexation, and acid-base neutralization reactions.

Although a variety of instrumental methods for detecting equivalence points are now available, it is frequently more convenient to add an indicator to the reaction mixture. An indicator is a substance that undergoes a distinct color change at or near the equivalence point. The point in the titration at which the color change occurs is called the **end point**. Obviously, the titration will be accurate only if the end point and the equivalence point coincide fairly closely. For this reason, the indicator used in a titration must be selected carefully. Fortunately, a large number of indicators are commercially available and finding the right one for a particular titration is not a difficult task.

Acids and Bases

Although several definitions of acids and bases may be given, the classical Arrhenius concept will suffice for this experiment. According to this concept, an acid is a substance that dissociates in water to produce hydrogen ions; a base is a substance that dissociates in water to produce hydroxide ions. The classical Arrhenius acid-base reaction is one in which an acid reacts with a base to form water (from the combination of hydrogen ions and hydroxide ions) and a salt. Such a reaction is called a **neutralization reaction**. The neutralization reaction of acetic acid, the primary ingredient of vinegar, and sodium hydroxide is shown in Equation 8P.1.

$$HC_2H_3O_2\ (aq) + NaOH\ (aq) \rightarrow H_2O\ (l) + NaC_2H_3O_2\ (aq) \qquad (8P.1)$$

Because the mole ratio of acetic acid to sodium hydroxide is 1:1, the number of moles of acid present in the sample is equal to the number of moles of base that must be added to reach the equivalence point of the titration (see Equation 8P.2).

$$M_b \times V_b = \text{moles base} = \text{moles acid} = M_a \times V_b \qquad (8P.2)$$

Indicators

Acid-base indicators are weak acids or bases that have different colors when in their dissociated and undissociated forms, respectively. The dissociation of the indicator HIn may be represented as shown in Equation 8P.3.

$$HIn + H_2O \rightleftharpoons H_3O^+ + In^- \qquad (8P.3)$$

In acid solutions, the indicator exists predominantly as HIn. In basic solutions, it is present mainly as In^- ions, which impart a different color to the solution. Phenolphthalein is an indicator that is used frequently in acid-base titrations. It is colorless as long as the titration mixture contains an excess of acid and turns pink when the amount of base added slightly exceeds the amount of acid originally present. Thus, the end point will come shortly after the equivalence point. The conditions that cause phenolphthalein to change from its colorless (HIn) form to its colored (In^-) form are such that the difference between the end point and the equivalence point in a carefully performed titration is negligible.

Preparation of NaOH *Solution*

The general procedure for preparing carbonate-free NaOH solution involves decanting (pouring off) clear liquid from a 50% aqueous NaOH solution and then diluting this with freshly boiled and cooled water. This precaution is needed because atmospheric carbon dioxide tends to react with sodium hydroxide to form sodium carbonate. Because sodium carbonate has a very low solubility in 50% NaOH solution, any sodium carbonate present as an impurity in the solid NaOH used to prepare the solution (or formed by reaction of the solution of NaOH with carbon dioxide in the air) does not remain in solution. Rather, it slowly settles to the bottom of the solution as a white precipitate, from which the pure NaOH solution can be separated easily by decantation. The water used subsequently to dilute this solution should first be boiled to remove dissolved carbon dioxide.

Because of the very great hazard to the eyes presented by concentrated solutions of bases, you will not be given 50% NaOH. Rather, dilute (approximately 0.05 M), carbonate-free solution will be provided for your use. You will determine the exact concentration of NaOH in the approximately 0.05 M solution by titrating a known mass of potassium hydrogen phthalate. The reaction between these substances is shown in Equation 8P.4.

$$NaOH\ (aq) + KHC_8H_4O_4\ (aq) \rightarrow H_2O\ (l) + KNaC_8H_4O_4\ (aq) \qquad (8P.4)$$

Thus, 1 mole of NaOH reacts exactly with 1 mole (204.22 g) of potassium hydrogen phthalate ($KHC_8H_4O_4$, generally labeled KHP).

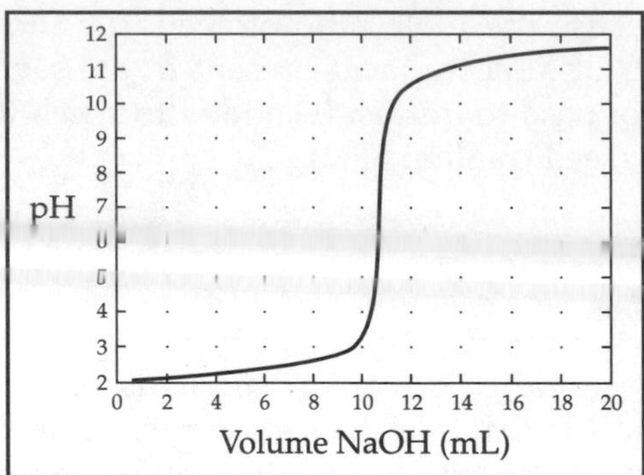

Figure 8P.1

Procedure

Standardization of NaOH Solution

Clean a buret and prepare it for use in the titration according to the directions provided in the Introduction. Rinse the buret with small amounts of distilled water. Then rinse it twice with small portions of the NaOH solution provided. Finally, fill the buret with this solution.

Accurately weigh, to the nearest tenth of a milligram, 0.2 to 0.4 g of the dry KHP and add it to a 125-mL Erlenmeyer flask. Add 25 mL of distilled water and 2 to 3 drops of phenolphthalein indicator solution. Swirl to dissolve the solid. Read the initial volume of NaOH solution in the buret to the nearest hundredth of a milliliter.

Plug the pH Sensor into Channel 1 of the CBL 2 interface. Use the link cable to connect the TI Graphing Calculator to the interface. Firmly press in the cable ends.

Use a utility clamp to suspend a pH Sensor on a ringstand as shown in Figure 8P.2. Position the pH Sensor in the HCl solution and adjust its position so that it is not struck by the stirring bar.

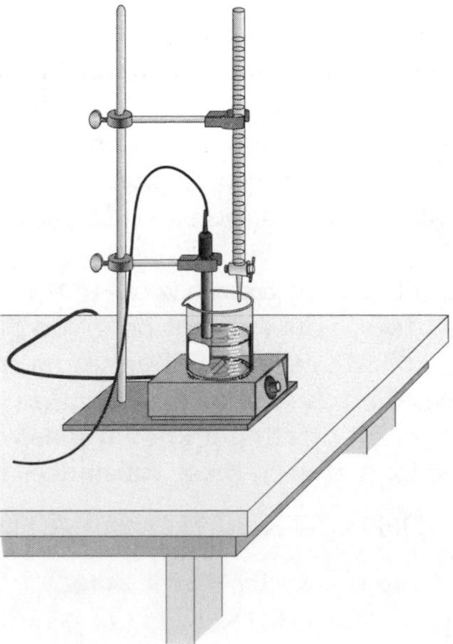

Figure 8P.2

Turn on the calculator and start the DataMate program. Press [CLEAR] to reset the program.

1. Set up the calculator and interface for the pH sensor.

   Select SETUP from the main screen.

   If CH 1 displays PH, proceed directly to Step 2. If it does not, continue with this step to set up your sensor manually.

   Press [ENTER] to select CH 1.

   Select PH from the SELECT SENSOR menu.

2. Set up the data-collection mode.

   To select MODE, press [▲] once and press [ENTER].

   Select EVENTS WITH ENTRY from the SELECT MODE menu.

   Select OK to return to the main screen.

You are now ready to perform the titration. This process goes faster if one person manipulates and reads the buret while another person operates the calculator and enters volumes.

Select START to begin data collection.

Before you begin to add any NaOH solution, press [ENTER] and type in "0" as the buret volume in mL. Press [ENTER] to save the first data pair for this experiment.

Add the next increment of NaOH titrant (enough to raise the pH about 0.15 units). When the pH stabilizes, press [ENTER] and enter the current buret reading (to the nearest 0.01 mL). You have now saved the second data pair for the experiment.

Continue adding NaOH solution in increments that raise the pH by about 0.15 units and enter the buret reading after each increment. When a pH value of approximately 3.5 is reached, change to a one-drop increment. Enter a new buret reading after each increment. Note: It is important that all increment volumes in this part of the titration be equal; that is, one-drop increments.

After a pH value of approximately 10 is reached, again add larger increments that raise the pH by about 0.15 pH units, and enter the buret level after each increment.

Continue adding NaOH solution until the pH value remains constant.

Press [STO▶] when you have finished collecting data.

Examine the data on the displayed graph to find the *equivalence point*—that is the largest increase in pH upon the addition of 1 drop of NaOH solution. As you move the cursor right or left on the displayed graph, the volume (X) and pH (Y) values of each data point are displayed below the graph. Go to the region of the graph with the largest increase in pH. Find the NaOH volume just *before* this jump. Record this value in the data table. Then record the NaOH volume *after* the drop producing the largest pH increase was added.

Titrate the KHP solution with the NaOH solution until a faint pink color, which does not disappear when the solution is mixed, is obtained. Read the final volume of NaOH solution in the buret to the nearest hundredth of a milliliter. Calculate the molarity of the NaOH solution. Repeat the standardization and average the results. If time permits, you may wish to perform a third standardization and use the criteria given in the Introduction to decide whether you should average the results of two or three trials.

Titration of an Unknown

Rinse a 10-mL pipet with small portions of distilled water; then, use the pipet to deliver a 10-mL sample of water to a pre-weighed Erlenmeyer flask. Determine the mass and temperature of the water delivered; find the actual volume of water delivered by the pipet.

Rinse the pipet with two small volumes of vinegar. Then, use the pipet to deliver a 10-mL sample of vinegar to a clean (but not necessarily dry) 100-mL volumetric flask. Add 50 mL of distilled water to the vinegar and swirl the flask carefully to mix its contents. Dilute the solution to the 100-mL mark with distilled water; invert the capped flask several times to promote further mixing of the diluted vinegar.

Rinse the pipet with small portions of your diluted vinegar solution, ultimately using the pipet to deliver a 10-mL sample of diluted vinegar to a 125-mL Erlenmeyer flask. Add 2 or 3 drops of phenolphthalein indicator and then titrate the acid with your 0.05 M NaOH solution until a faint, persistent pink color is obtained. Calculate the molarity of the acetic acid solution. Repeat the titration once or twice. Report the results of each titration and the average concentration of your diluted vinegar. Also report the concentration of the vinegar (prior to dilution).

Disposal of Reagents

Excess KHP should be placed in the containers used for solid waste. Solutions should be neutralized and diluted. They may then be flushed down the drain.

EXP
9

Job's Method for Determining the Stoichiometry of a Reaction

Laboratory Time Required

Three hours.

Special Equipment and Supplies

Analytical balance
Volumetric flask, 1-L
Storage bottle, 1-L
Graduated cylinders, 100-mL
Buret
Buret clamp
Water bath
Coffee cup calorimeters
Thermometer

1 M sodium hydroxide, NaOH
Citric acid, $C_6H_8O_7 \cdot H_2O$ (*s*)
Potassium hydrogen phthalate (KHP)
Phenolphthalein

Objective

In performing this experiment, students will standardize a sodium hydroxide solution via titration, prepare a standard solution, and determine the stoichiometric ratio of the neutralization reaction of sodium hydroxide and citric acid.

Safety

Bases, such as sodium hydroxide, can cause skin burns and are especially hazardous to the eyes.

Safety goggles are required for this lab.

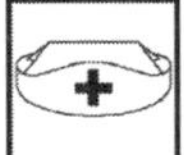

First Aid

Following skin contact with sodium hydroxide, wash the area thoroughly with water. If there has been eye contact, flush the eyes with water for at least 20 minutes and seek medical attention.

Preamble

Like many organic acids, citric acid contains some hydrogens that are acidic and others that are not. In this experiment, the reaction between citric acid and sodium hydroxide is studied and the stoichiometric ratio of the reactants is determined.

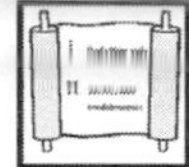

Principles

Job's method of continuous variation offers a relatively quick procedure for determining the stoichiometric ratio between two reactants. One form of Job's method involves the preparation of a series of mixtures in which the chemical amount of each reactant varies, although the total number of moles in each mixture remains the same. Thus, in the study used to prepare the graph shown in Figure 9.1a, 0.100 M solutions of reagent A were mixed with 0.100 M solutions of reagent B. The volumes of reagent A used varied from 0.0 mL to 20.0 mL, while the volumes of reagent B used varied from 20.0 mL to 0.0 mL. One mixture studied contained 2.00 mmols of A; another contained 0.80 mmols of A and 1.20 mmols of B; a third contained 1.60 mmols of A and 0.40 mmols of B. In each case, the total number of mmols was 2.00. Of course, in some of the mixtures, A was in excess. In others, B was in excess and A was the limiting reagent. In the original study, only even volumes (2.00 mL, 4.00 mL, etc.) were used. However, because most mixtures studied contained an excess of A, it was difficult to determine how to best fit the points that were marked by an excess of B. Therefore, additional mixtures were prepared (one containing 0.10 mmol of A and 1.90 mmol of B and another containing 0.30 mmol of A and 1.70 mmol of B) and data obtained from studying these mixtures was added to the graph. (See Figure 9.1b.)

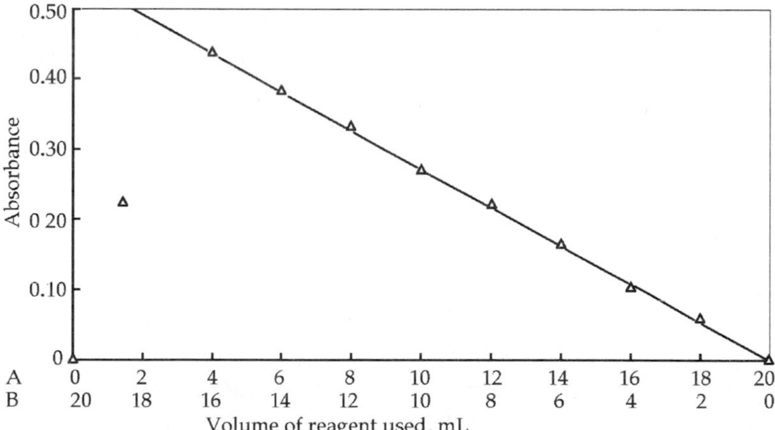

Figure 9.1a Graph of Absorbance versus Composition for Job's Method Study of A and B.

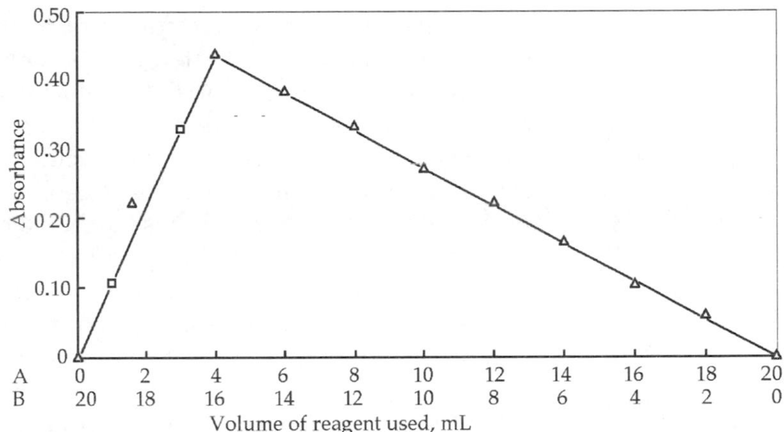

Figure 9.1b Absorbance versus Composition for Job's Method Study of A and B, with added data.

Job's method utilizes some factor (color change, temperature change, volume change) that is maximized when the reactants are mixed in essentially their stoichiometric ratio (and, consequently, there is no excess of either reagent). The graphs in Figure 9.1 were prepared by measuring the absorbances of the various mixtures of A and B at a wavelength of light that was absorbed by the product of their reaction. Note that the x-axis is labeled with both the amount of A and the amount of B in each mixture. The maximum in absorbance is observed to occur in a mixture containing 0.40 mmols of A and 1.60 mmols of B, indicating that the stoichiometric ratio is 1A : 4B. In the case discussed here, one of the mixtures had the exact stoichiometric ratio of the reaction. In general, there is no guarantee that there will be a mixture in which neither reagent is in excess. Thus, the point at which the two "best fit" lines cross may very well not be one of the experimental data points.

RESULTS

Procedure

The sodium hydroxide provided for this experiment should be close to 1 M. Standardize this solution by titrating it against 6 g samples of KHP (which have been weighed to the nearest 0.1 mg on the analytical balance). Use phenolphthalein as the indicator. Review Experiment 8, if necessary.

Once the molarity of the sodium hydroxide solution has been determined, prepare 1.00 L of a solution of citric acid of exactly the same concentration. Be sure to mix the solution well. Once the solution of citric acid has been prepared, place the volumetric flask containing the solution and the bottle containing the sodium hydroxide solution in a water bath containing room-temperature water. Allow the solutions to equilibrate in the water bath for 10 to 15 minutes.

Procedure in a Nutshell

Standardize a solution of sodium hydroxide. Prepare 1.00 L of a solution of citric acid that has the same molarity as the sodium hydroxide solution. Place the two solutions in a water bath to ensure that both are at the same temperature. Measure the temperature change as various volumes of the two solutions are mixed. Plot the temperature change versus solution composition. If the study of solutions specified in the Procedure section does not yield sufficient information, prepare additional solutions until the data permits an unambiguous determination of the stoichiometric ratio of sodium hydroxide to citric acid in the neutralization reaction.

Prepare a series of coffee cup calorimeters by nesting sets of two cups together and placing the nested cups in empty beakers to enhance the stability of the calorimeter. Label the calorimeters with the volumes listed in Table 9.1. Use 100-mL graduated cylinders to measure the volumes of sodium hydroxide and of citric acid specified. Begin each run by placing the acid to be used in that run in a coffee cup calorimeter. Monitor the temperature of the acid. When it has reached a constant value, quickly add the amount of sodium hydroxide specified for the run. Stir this reaction mixture and monitor its temperature at 30-second intervals until the temperature begins to fall. Repeat this procedure until all mixtures listed in Table 9.1 have been studied. Take the difference between the maximum temperature observed for each mixture and its starting temperature as Δt. Plot Δt versus composition, as absorbance was plotted against composition in Figure 9.1. ***Do this before you dispose of your solutions.*** If the data collected permit you to do so, determine the stoichiometric ratio for the neutralization of citric acid with sodium hydroxide. If you feel that you need to study additional solutions before you can make the determination of the stoichiometric ratio, prepare the mixtures you wish to study, measure Δt for each of the additional mixtures, and plot these Δt values along with the other ones on your graph.

Table 9.1 Volumes of sodium hydroxide and citric acid solutions to be used in determining the stoichiometric ratio of their neutralization reaction

Mixture	*Volume of* NaOH *(aq), mL*	*Volume of citric acid (aq), mL*
1	0.0	100.0
2	10.0	90.0
3	20.0	80.0
4	30.0	70.0
5	40.0	60.0
6	50.0	50.0
7	60.0	40.0
8	70.0	30.0
9	80.0	20.0
10	90.0	10.0
11	100.0	0.0

name | section | date

Pre-Lab Exercises for Experiment 9

These exercises are to be completed after you have read the experiment but before you come to the laboratory to perform it.

Suppose you used Job's method to find the stoichiometry of the reaction:

$$a\, H_2\ (g) + b\, O_2\ (g) \rightarrow \text{Product}\ (l)$$

The results of the study of seven mixtures are shown below.

	Moles Initially Present			
Mixture	H_2	O_2	*Moles of Product*	*Moles of Excess Reactant*
1	1.0	7.0	1.0	6.5
2	2.0	6.0	2.0	5.0
3	3.0	5.0	3.0	3.5
4	4.0	4.0	4.0	2.0
5	5.0	3.0	5.0	0.5
6	6.0	2.0	4.0	2.0
7	7.0	1.0	2.0	5.0

1. Use the form shown on the following page to plot the yield of product (in moles) versus the composition for each of the seven mixtures. Draw two straight lines through the experimental points and, from their intersection, deduce the optimum ratio of $H_2(g)$ to $O_2(g)$. Find the formula of the product and write a balanced chemical equation for the reaction.

2. Repeat the operation performed in answer to Pre-Lab Exercise 1, using the number of moles of unreacted gas (rather than the moles of product) to label the *y*-axis.

3. Do your answers for Pre-Lab Exercises 1 and 2 agree? Explain your response, briefly.

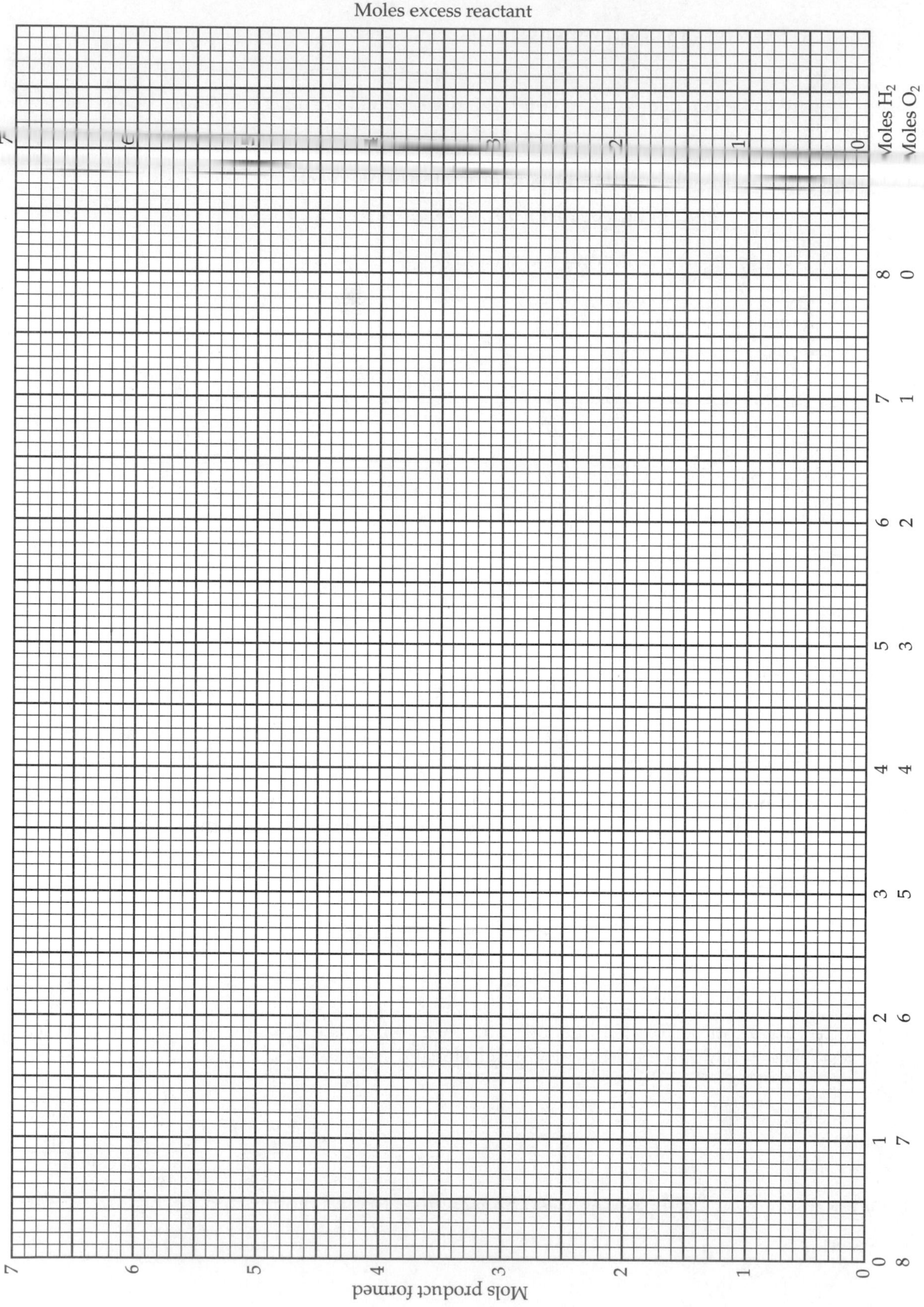
Moles excess reactant
Mols product formed
Moles H_2
Moles O_2
0 1 2 3 4 5 6 7
8 7 6 5 4 3 2 1 0
0 1 2 3 4 5 6 7 8

name ______________________ section ______________________ date ____________

Summary Report on Experiment 9

Standardization of Sodium Hydroxide

Mass of container plus KHP ____________

Mass of container ____________

Mass of KHP ____________

Final buret reading ____________

Initial buret reading ____________

Volume of NaOH used ____________

Molarity of NaOH ____________

Preparation of citric acid solution

Mass of citric acid required to prepare 1.00 L of solution ____________

Mass of container plus citric acid ____________

Mass of container ____________

Mass of citric acid ____________

Determining the value of Δt

Mixture	____________	Initial temperature	____________
Volume NaOH	____________	Volume citric acid	____________
Temperature	____________	Temperature	____________
Temperature	____________	Temperature	____________
Temperature	____________	Temperature	____________
Temperature	____________	Temperature	____________
Temperature	____________	Temperature	____________
Temperature	____________	Temperature	____________
		Δt =	____________

Mixture	______	Initial temperature	______
Volume NaOH	______	Volume citric acid	______
Temperature	______	Temperature	______
Temperature	______	Temperature	______
Temperature	______	Temperature	______
Temperature	______	Temperature	______
Temperature	______	Temperature	______
Temperature	______	Temperature	______
		$\Delta t =$	______

Mixture	______	Initial temperature	______
Volume NaOH	______	Volume citric acid	______
Temperature	______	Temperature	______
Temperature	______	Temperature	______
Temperature	______	Temperature	______
Temperature	______	Temperature	______
Temperature	______	Temperature	______
Temperature	______	Temperature	______
		$\Delta t =$	______

Mixture	______	Initial temperature	______
Volume NaOH	______	Volume citric acid	______
Temperature	______	Temperature	______
Temperature	______	Temperature	______
Temperature	______	Temperature	______
Temperature	______	Temperature	______
Temperature	______	Temperature	______
Temperature	______	Temperature	______
		$\Delta t =$	______

name ______________________ section ______________________ date __________

Mixture	______	Initial temperature	______
Volume NaOH	______	Volume citric acid	______
Temperature	______	Temperature	______
Temperature	______	Temperature	______
Temperature	______	Temperature	______
Temperature	______	Temperature	______
Temperature	______	Temperature	______
Temperature	______	Temperature	______
		$\Delta t =$	______

Mixture	______	Initial temperature	______
Volume NaOH	______	Volume citric acid	______
Temperature	______	Temperature	______
Temperature	______	Temperature	______
Temperature	______	Temperature	______
Temperature	______	Temperature	______
Temperature	______	Temperature	______
Temperature	______	Temperature	______
		$\Delta t =$	______

Mixture	______	Initial temperature	______
Volume NaOH	______	Volume citric acid	______
Temperature	______	Temperature	______
Temperature	______	Temperature	______
Temperature	______	Temperature	______
Temperature	______	Temperature	______
Temperature	______	Temperature	______
Temperature	______	Temperature	______
		$\Delta t =$	______

Mixture	___________	Initial temperature	___________
Volume NaOH	___________	Volume citric acid	___________
Temperature	___________	Temperature	___________
Temperature	___________	Temperature	___________
Temperature	___________	Temperature	___________
Temperature	___________	Temperature	___________
Temperature	___________	Temperature	___________
Temperature	___________	Temperature	___________
		$\Delta t =$	___________

Mixture	___________	Initial temperature	___________
Volume NaOH	___________	Volume citric acid	___________
Temperature	___________	Temperature	___________
Temperature	___________	Temperature	___________
Temperature	___________	Temperature	___________
Temperature	___________	Temperature	___________
Temperature	___________	Temperature	___________
Temperature	___________	Temperature	___________
		$\Delta t =$	___________

Mixture	___________	Initial temperature	___________
Volume NaOH	___________	Volume citric acid	___________
Temperature	___________	Temperature	___________
Temperature	___________	Temperature	___________
Temperature	___________	Temperature	___________
Temperature	___________	Temperature	___________
Temperature	___________	Temperature	___________
Temperature	___________	Temperature	___________
		$\Delta t =$	___________

name	section	date

Mixture	______	Initial temperature	______
Volume NaOH	______	Volume citric acid	______
Temperature	______	Temperature	______
Temperature	______	Temperature	______
Temperature	______	Temperature	______
Temperature	______	Temperature	______
Temperature	______	Temperature	______
Temperature	______	Temperature	______
		Δt =	______

Plot the data you have acquired thus far. If you do not feel that these data will provide you with an unambiguous value for the stoichiometric ratio of sodium hydroxide to citric acid, decide on the composition of additional mixtures, which you should prepare. Use the portion of the Summary Report Sheet given below to record this additional data.

Mixture	______	Initial temperature	______
Volume NaOH	______	Volume citric acid	______
Temperature	______	Temperature	______
Temperature	______	Temperature	______
Temperature	______	Temperature	______
Temperature	______	Temperature	______
Temperature	______	Temperature	______
Temperature	______	Temperature	______
		Δt =	______

Mixture	___________	Initial temperature	___________
Volume NaOH	___________	Volume citric acid	___________
Temperature	___________	Temperature	___________
Temperature	___________	Temperature	___________
Temperature	___________	Temperature	___________
Temperature	___________	Temperature	___________
Temperature	___________	Temperature	___________
Temperature	___________	Temperature	___________
		$\Delta t =$	___________

Mixture	___________	Initial temperature	___________
Volume NaOH	___________	Volume citric acid	___________
Temperature	___________	Temperature	___________
Temperature	___________	Temperature	___________
Temperature	___________	Temperature	___________
Temperature	___________	Temperature	___________
Temperature	___________	Temperature	___________
Temperature	___________	Temperature	___________
		$\Delta t =$	___________

Mixture	___________	Initial temperature	___________
Volume NaOH	___________	Volume citric acid	___________
Temperature	___________	Temperature	___________
Temperature	___________	Temperature	___________
Temperature	___________	Temperature	___________
Temperature	___________	Temperature	___________
Temperature	___________	Temperature	___________
Temperature	___________	Temperature	___________
		$\Delta t =$	___________

name ______ section ______ date ______

Mixture	______	Initial temperature	______
Volume NaOH	______	Volume citric acid	______
Temperature	______	Temperature	______
Temperature	______	Temperature	______
Temperature	______	Temperature	______
Temperature	______	Temperature	______
Temperature	______	Temperature	______
Temperature	______	Temperature	______
		$\Delta t =$	______

Mixture	______	Initial temperature	______
Volume NaOH	______	Volume citric acid	______
Temperature	______	Temperature	______
Temperature	______	Temperature	______
Temperature	______	Temperature	______
Temperature	______	Temperature	______
Temperature	______	Temperature	______
Temperature	______	Temperature	______
		$\Delta t =$	______

Stoichiometric ratio of NaOH to citric acid ______

Attach your plot of Δt versus composition of the sodium hydroxide/citric acid mixtures.

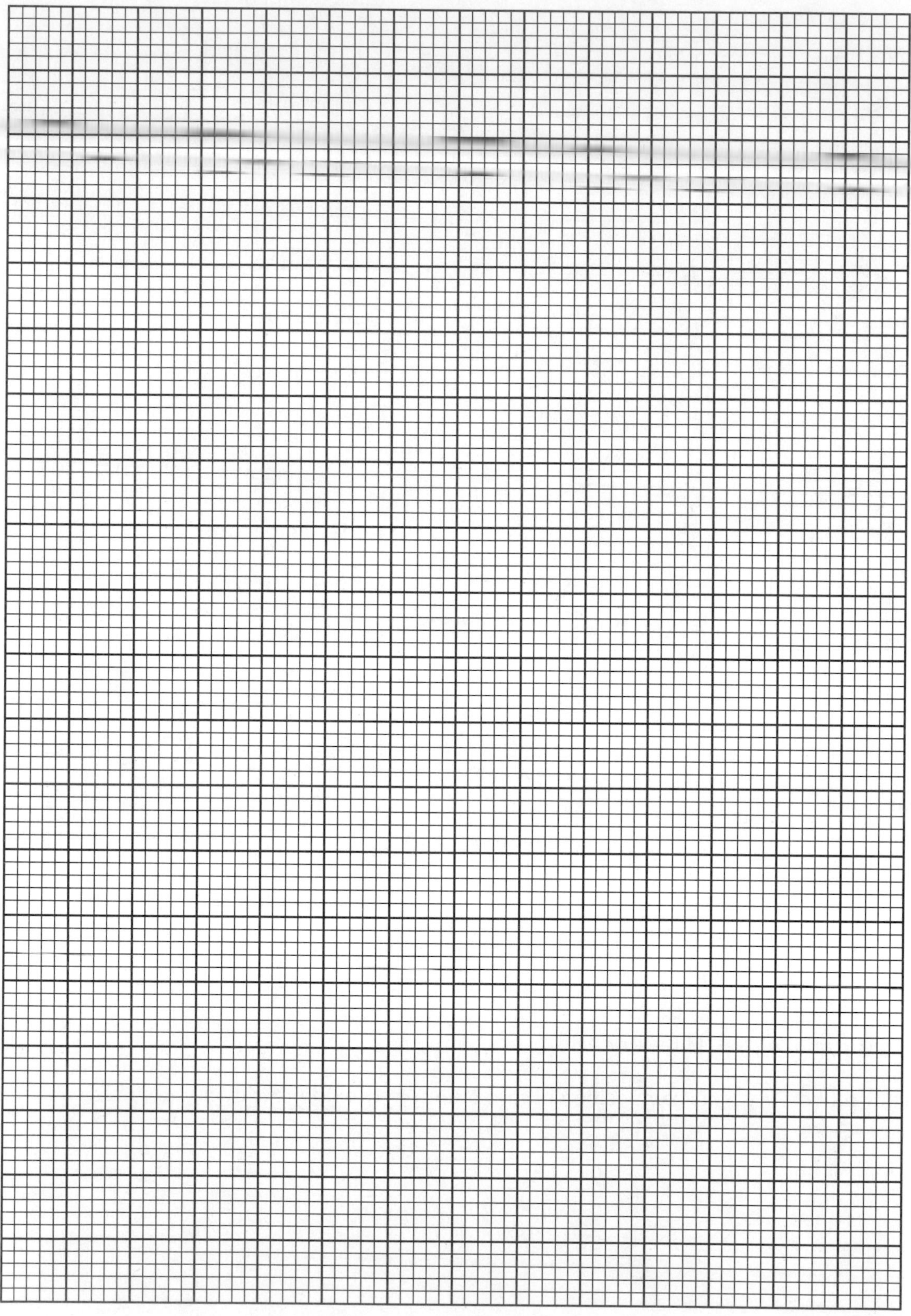

The Synthesis of Cobalt Oxalate Hydrate

EXP
10

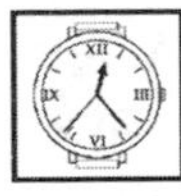

Laboratory Time Required

One hour. May conveniently be combined with Experiment 11.

Special Equipment and Supplies

Analytical balance
Centrifuge
Oxalic acid dihydrate, $H_2C_2O_4 \cdot 2H_2O$
Cobalt(II) chloride hexahydrate, $CoCl_2 \cdot 6H_2O$
Litmus paper
Concentrated ammonia, 15 M NH_3 (*aq*)

Objective

In performing this experiment, students will learn to work with hydrated compounds and to weigh out specified masses quickly and accurately. They will prepare a compound, which they may subsequently analyze for cobalt content, oxalate content, and water content.

Safety

All of the chemicals used in this experiment are hazardous! Avoid getting the solutions on the skin, in the eyes, or in the mouth. Use the pipet bulb.

Oxalic acid is a **poison**.
Cobalt chloride is **toxic**.

Aqueous ammonia is **caustic**. Its vapor causes eye irritation.

Because of the toxic and caustic nature of the chemicals used, safety goggles are **mandatory**, and laboratory aprons are strongly advised.

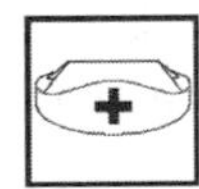

First Aid

If you get oxalic acid on your skin, flush the area with water. Then flush the skin with aqueous sodium bicarbonate.

If you ingest oxalic acid, drink a large quantity of water, followed by milk or milk of magnesia. See a doctor.

If you have contact with aqueous ammonia, flush the area with water.

Use the eyewash fountain promptly to remove ammonia from the eyes. See a doctor if ammonia has gotten into your eyes.

Preamble

In this experiment, you are to synthesize a substance that will be given the tentative formula $Co_a(C_2O_4)_b \cdot cH_2O$. You may subsequently analyze the compound in Experiment 12, or your instructor may give you the values of *a*, *b*, and *c*.

Principles

Hydrates

Compounds that contain water may have it present either in a variable or in a definite mass percentage. If the water content is variable, the substance may be merely wet; or the water may occupy channels, as in a zeolite; or a mixture of hydrates may be present. If the substance has a definite, nonvariable water content, it is called a **hydrate**, and the water molecules are found to occupy definite sites in the crystal. A few common hydrates and their formulas are: borax ($Na_2B_4O_7 \cdot 10H_2O$), plaster of Paris ($CaSO_4 \cdot \frac{1}{2}H_2O$), epsom salt ($MgSO_4 \cdot 7H_2O$), and blue copper sulfate pentahydrate ($CuSO_4 \cdot 5H_2O$).

Under appropriate conditions, most hydrated salts can be dehydrated to form either a lower hydrate or the anhydrous salt. For example, when the mineral gypsum is heated no hotter than about 160°C, it dehydrates to form plaster of Paris as shown in Equation 10.1.

$$CaSO_4 \cdot 2H_2O \rightleftharpoons CaSO_4 \cdot \tfrac{1}{2}H_2O + \tfrac{3}{2}H_2O \qquad (10.1)$$

This reaction is reversible, and the setting of plaster of Paris may be attributed to its rehydration to form crystalline gypsum. The equilibrium between the hydrated and the anhydrous forms of the salt is influenced both by the temperature of the system and by the relative humidity. Salts that can be reversibly hydrated are frequently used to control the amount of moisture in the air. Anhydrous salts that are commonly used as desiccants (drying agents) include $CaSO_4$, $CaCl_2$, and $Mg(ClO_4)_2$.

Calculation of Percent Yield

If stoichiometric quantities of chemicals are mixed together when a compound is being synthesized, then all reactants will be used up at the same time. No excess unreacted starting material will remain at the completion of the reaction. On the other hand, if the reactants are mixed in a nonstoichiometric ratio, then one reactant will be used up before the others. Because no more product can be formed when the supply of one of the reactants has been exhausted, the first reactant to be consumed limits the amount of product that can be formed and is designated as the **limiting reactant** or **limiting reagent**.

The amount of product formed when the limiting reagent has been completely consumed is called the **theoretical yield of the reaction**. In practice, the actual yield of product is frequently lower than this theoretical maximum (because side reactions occur, or product is lost in collection, or because the product is unstable). The percent yield of the reaction is a measure of success of the reaction. It is defined in Equation 10.2.

$$\text{Percent yield} = \frac{\text{actual yield}}{\text{theoretical yield}} \times 100\% \qquad (10.2)$$

If you will not be performing Experiment 12, your instructor will supply you with the values of *a*, *b*, and *c* in the formula $Co_a(C_2O_4)_b \cdot cH_2O$.

You should then be able to calculate your percent yield of cobalt oxalate hydrate based on the amounts of the starting materials you used in this synthesis. If you will be doing Experiment 12, you will be able to calculate the percent yield after you have performed your analyses and found the empirical formula of the compound.

Procedure

Procedure in a Nutshell

Dissolve a known mass of oxalic acid dihydrate in water and add concentrated ammonia. Dissolve a known mass of cobalt(II) chloride hexahydrate in another portion of water. Add the second solution to the first, dropwise, while stirring. Cool the mixture of solutions in an ice bath. Collect the resulting precipitate by gravity filtration. Allow the collected precipitate to dry.

Synthesis of Cobalt Oxalate Hydrate

Place 100 mL of distilled water in a 250-mL (or 400-mL) beaker. Add 1.26 g of oxalic acid dihydrate ($H_2C_2O_4 \cdot 2H_2O$) and 1 mL of concentrated ammonia. Stir the mixture until the solid has dissolved completely.

Dissolve 2.34 g of cobalt chloride hexahydrate ($CoCl_2 \cdot 6H_2O$) in 100 mL of water in an Erlenmeyer flask. While stirring the oxalic acid solution constantly, add the cobalt chloride solution drop by drop. Let the mixture cool in an ice bath. A precipitate will form slowly.

After the precipitate has had a chance to settle, collect it by gravity filtration. Wash the collected solid sparingly with cold water. Allow the water to drain from the collected solid. Then transfer the filter paper and precipitate to a paper towel so that excess moisture can be absorbed. Before you leave the lab, place your precipitate (still in its filter paper cone) in a beaker and leave it in your desk to air dry for at least three days.

After your precipitate has dried thoroughly, weigh it on the analytical balance. Report your yield of product. If your instructor has given you the values of *a*, *b*, and *c*, determine the limiting reagent in your synthesis and report your percent yield. If you will be performing Experiment 12, save your product for analysis.

Disposal of Reagents

The filtrate from the synthesis of cobalt oxalate should not be poured down the drain before it has been tested for the complete removal of Co^{2+} ions. About 1 millimole or 60 mg of Co^{2+} typically remains in solution. To test, add 3 M NaOH by drops until the solution tests basic to litmus paper. If no precipitate forms, dilute and discard the solution. If a precipitate forms, allow it to settle. Then decant the clear solution into another beaker for dilution and disposal. Pour the $Co(OH)_2$ slurry into the designated collection bottle. At a later time, this may be filtered, ignited to Co_3O_4, and bottled for disposal in a hazardous waste landfill.

Leftover oxalic acid may be decomposed by acidification with H_2SO_4, followed by oxidation of the oxalate to CO_2 with $KMnO_4$. This may be accomplished by following the procedure given in Experiment 12 for the determination of oxalate. The resulting solution should be made just basic to precipitate $Mn(OH)_2$, which should then be placed in a collection bottle labeled $Mn(OH)_2$. This product can be subsequently ignited to produce Mn_3O_4, which may be placed in a hazardous waste landfill site.

Excess NH_3 solutions should be neutralized, diluted, and poured down the drain.

The cobalt oxalate hydrate product should be saved for analysis in Experiment 12 or, if Experiment 12 is not to be performed, placed in a separate collection bottle. The compound can be ignited to Co_3O_4 when convenient and then deposited in a hazardous waste landfill site.

name section date

Pre-Lab Exercises for Experiment 10

These exercises are to be done after you have read the experiment but before you come to the laboratory to perform it.

The equation given below refers to the reaction of ammonium chloride and barium hydroxide octahydrate.

$$2NH_4Cl(s) + Ba(OH)_2 \cdot 8H_2O(s) \rightarrow 2NH_3(g) + BaCl_2(aq) + 10H_2O(l)$$

In a particular experiment, 150.0 g of ammonium chloride were reacted with 290.0 g of barium hydroxide octahydrate, producing 157.2 g of water.

1. Find the limiting reagent in this situation.

2. Calculate the theoretical yield of water.

3. Calculate the percent yield of water.

4. What mass of anhydrous barium hydroxide contains the same number of moles of Ba^{2+} ions as 31.5 g of barium hydroxide octahydrate?

name ______ section ______ date ______

Summary Report on Experiment 10

Mass of container plus oxalic acid dihydrate	______
Mass of container	______
Mass of oxalic acid dihydrate	______
Mass of container plus cobalt chloride hexahydrate	______
Mass of container	______
Mass of cobalt chloride hexahydrate	______
Mass of product	______
Values of *a*, *b*, *c* (if supplied)	______
Limiting reagent	______
Percent yield	______

EXP
11

The Synthesis of a Nitrite Complex

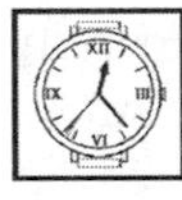

Laboratory Time Required

One and one-half hours. May be combined with Experiment 10.

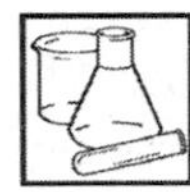

Special Equipment and Supplies

Analytical balance
Desiccator
Buchner funnel and vacuum flask
Ice bath

Transition metal chlorides

Copper(II) chloride, $CuCl_2 \cdot xH_2O$
Cobalt(II) chloride, $CoCl_3 \cdot xH_2O$
Nickel(II) chloride, $NiCl_2 \cdot xH_2O$

Alkaline earth metal chlorides

Calcium chloride, $CaCl_2 \cdot xH_2O$
Strontium chloride, $SrCl_2 \cdot xH_2O$
Barium chloride, $BaCl_2 \cdot xH_2O$
Potassium nitrite, KNO_2
Acetone, C_3H_6O
Anhydrous calcium chloride, $CaCl_2$, as a desiccant

Objective

In performing this experiment, students will learn to work with hydrated compounds and to weigh out calculated masses quickly and accurately. They will prepare a compound that they may subsequently analyze.

Safety

The KNO_2 is toxic (when ingested).
The salts of Co^{2+}, Ni^{2+}, and Sr^{2+} are toxic.

KNO_2 is also a strong oxidizing agent. Take care to avoid contacting KNO_2 with your skin, mouth, or eyes.

You should also dispose of excess KNO_2 in a way such that it does not come in contact with reducing agents.

Acetone is flammable. There should be no flames in the laboratory when acetone is in use.

Because of the toxic and caustic nature of the chemicals used in this lab, safety goggles are **mandatory**, and lab aprons are strongly advised.

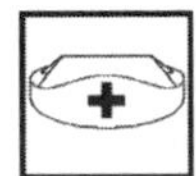

First Aid

First aid for exposure to any of these chemicals involves thorough flushing of the skin or eyes with water. Chemicals may be removed from the stomach by drinking large amounts of water and inducing vomiting.

Preamble

In this experiment, you will synthesize one of a group of compounds with the general formula $K_2MM'(NO_2)_6$, where M is an alkaline earth metal ion and M' is a divalent transition metal ion. These compounds are interesting because they illustrate the general properties of complex ions or coordination compounds and also because a large number of compounds of the same general formula can be prepared by varying the identities of M and M'.

Principles

Complex Ions

The structure of compound salts, such as $K_2BaCu(NO_2)_6$, was a question of great interest to inorganic chemists during the latter part of the nineteenth century. It was difficult to understand why compounds such as KNO_2, $Ba(NO_2)_2$, and $Cu(NO_2)_2$, which were perfectly stable in themselves, should combine further to form compound salts. The work of Alfred Werner and the concept of coordinate covalent bonding went a long way in solving the puzzle.

Werner postulated that compound salts contain complex ions, which are formed by the association of a metal ion with one or more molecular or ionic species, known as **ligands** or **complexing agents.** A familiar example is the $Ag(S_2O_3)_2^{3-}$ complex ion formed in the fixing process in photography by the reaction of hypo ($Na_2S_2O_3$ solution) with unexposed silver halide. Another familiar example is the deep blue $Cu(NH_3)_4^{2+}$ ion formed by the addition of ammonia to aqueous solutions containing Cu^{2+} ions. In countless other examples, complex ions play an important role in a technical process or an analytical procedure. In addition, many compounds of biochemical importance, such as chlorophyll and hemoglobin, contain complex ions.

In the compounds to be synthesized in this experiment, the complex ion is $M'(NO_2)_6^{4-}$, where M' is a transition metal. Combined with certain nontransition metals and potassium ions, these complexes are incorporated into compounds that tend to have low solubilities in water. Hence they may be isolated from aqueous solution.

The bonding in complex ions varies from predominantly ionic to predominantly covalent, depending on the nature of the metal ion and ligand involved. The transition metal ions, in particular, show a strong tendency to form complexes with both negative ions and molecules serving as ligands. This results from the relatively small size and large charge of these metal ions as well as the presence of a partially filled set of d orbitals, which may participate in covalent bonding. Owing to their larger size and lower charge, the representative metal ions of group I and group II in general form less-stable complexes than do the transition metal ions and do so with a more restricted group of ligands.

The number of bonding positions around the central metal ion that are occupied by nearest ligand atoms is known as the **coordination number** of the metal ion. The coordination number of M' in $M'(NO_2)_6^{4-}$ is six, with the $[O\text{-}N\text{-}O]^-$ ions arranged so that the six nitrogen atoms surround the metal ion in an octahedral arrangement. This coordination number is a common one for transition metal ions, although coordination numbers of two through twelve are known to occur. Examples of complexes illustrating coordination numbers two, four, six, and eight are shown in Figure 11.1.

$[H_3N-Ag-NH_3]^+$

Diamminesilver (I) ion
Linear

Coordination number two

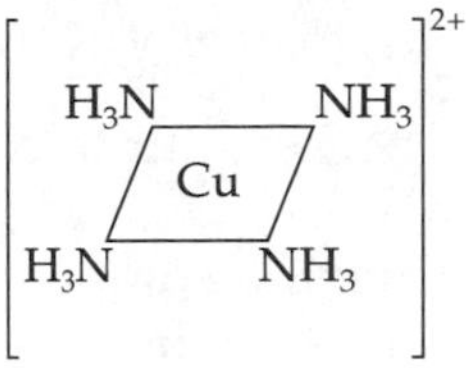

Tetraaminecopper (II) ion
Planar

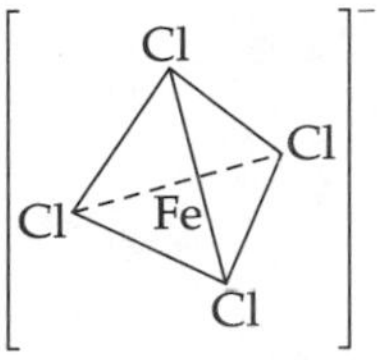

Tetrachloroferrate (III) ion
Tetrahedral

Coordination number four

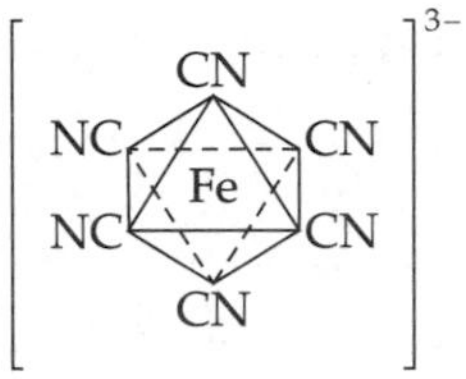

Hexacyanoferrate (III) ion
Octahedral

Coordination number six

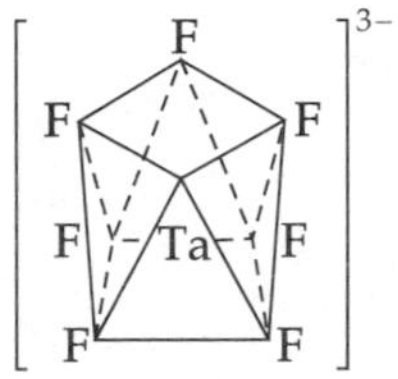

Octafluorotantalate (V) ion
Archimedean antiprism

Coordination number eight

Figure 11.1 Coordination numbers two, four, six, and eight

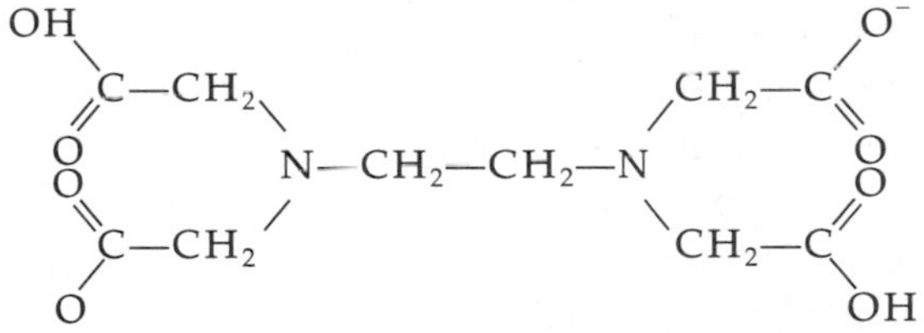

Figure 11.2 EDTA, a multidentant ligand

Note that the characteristic geometry for coordination number six is octahedral, whereas, for coordination number four, both tetrahedral and square planar geometries are observed in different complexes. Ligands such as Cl^-, NH_3, and H_2O can occupy only one coordination position around the central metal ion and are therefore known as **unidentate ligands. Multidentate ligands** contain two or more atoms, such as N, O, or S, which have unshared electron pairs that may be used to bond to a metal ion. These atoms are incorporated into a chain structure so that the ligand has several points of attachment to the metal ion. Ethylenediaminetetraacetic acid (EDTA, illustrated in Figure 11.2) is an example of a multidentate ligand. It is **hexadentate,** meaning that it can bind to a metal ion in six places. EDTA forms very stable complexes with a large number of metal ions and is a very useful reagent for analytical chemistry.

Preparation of $K_2MM'(NO_2)_6$ Compounds

The hexanitrite complexes of the divalent transition metal ions (such as Fe^{2+}, Co^{2+}, Ni^{2+}, or Cu^{2+}) are easily prepared by adding potassium nitrite to a solution of the desired transition metal ion in water. If the solution also contains a divalent nontransition metal ion (such as Ca^{2+}, Sr^{2+}, Ba^{2+}, or Pb^{2+}), the compound $K_2MM'(NO_2)_6$ precipitates as microcrystalline particles. The solid can be filtered and dried; the dried compound is stable in air. The solubility in water, however, depends on the identities of *M* and *M'*. Therefore the general procedure given here affords a reasonable yield of the solid product only for certain combinations of these ions. The recommended combinations are given in Table 11.1.

Procedure

Procedure in a Nutshell

Weigh out 0.005 mol of a chloride salt of each of your assigned metal ions. Dissolve both samples in approximately 5 mL of water. Weigh out 0.05 mol of potassium nitrite and dissolve it in approximately 5 mL of water. Add the potassium nitrite solution to the solution of metal ions, slowly, with stirring. Cool the mixture and transfer most of the liquid to a Buchner funnel. Wash the solid with a cold solution of potassium nitrite and use vacuum filtration to remove the liquid. Use acetone to dry the solid remaining in the Buchner funnel.

Obtain an assignment of transition metal and nontransition metal from your instructor. Weigh out 0.005 mole of the designated nontransition metal chloride and 0.005 mole of the designated transition metal chloride. (Remember to consider the numbers of waters of hydration in calculating the desired masses of the assigned salts.) Dissolve the two in a minimum volume (about 5 mL total) of distilled water, with low heat only if needed to dissolve the salts. Weigh out 0.05 mole of KNO_2, dissolve it in 5 mL of water, and then slowly add this solution, while stirring, to the solution of metal chlorides. Cool the mixture in an ice bath. Allow the solid to settle, then decant the supernatant liquid into a filter funnel, leaving most of the solid in the original beaker where it can be washed more efficiently. Add a 3-mL portion of cold water containing a small amount of KNO_2 to the solid, stir well, and pour the mixture onto the filter to collect the product. When the water has all drained out of the filter, pour 5 mL of acetone over the solid to help dry it. Pour a second 5-mL portion of acetone over the solid after the first has drained through the filter. Complete the drying of your compound by drawing a gentle stream of air through the filter using an aspirator. If no aspirator is available, open the filter paper and spread out the solid to dry in the air. The product is dry when it is powdery and has no odor of acetone.

Table 11.1 Recommended Combinations of Transition Metal Ions and Nontransition Metal Ions to Prepare $K_2MM'(NO_2)_6$ Compounds

Transition Metal Ion, M'	*Nontransition Metal Ions, M*
Ni^{2+}	Ca^{2+}, Sr^{2+}, or Ba^{2+}
Co^{2+}	Sr^{2+} or Ba^{2+}
Cu^{2+}	Sr^{2+}

After drying the solid, transfer it to a weighed bottle and reweigh. If your product loses weight during weighing, give the acetone more time to evaporate, then weigh the product again.

Disposal of Reagents

Although several of the chemicals used are hazardous, the small amounts remaining in solutions resulting from the synthetic procedure make it impractical to isolate the ions as insoluble salts. Consequently, the recommended disposal procedure for all solutions in this experiment is to dilute them tenfold and flush them down the drain with water. If you are to analyze your compound in Experiment 14, store it in a desiccator, over calcium chloride.

Questions

1. Determine which of the reactants used in your synthesis was the limiting reagent.

2. Calculate your percent yield of product.

name | section | date

Pre-Lab Exercises for Experiment 11

These exercises are to be completed after you have read the experiment but before you come to the laboratory to perform it.

1. Calculate the mass of potassium nitrite to be used in this experiment.

2. Robin Student calculated the mass of $CaCl_2 \cdot 6H_2O$ to be used in the synthesis of $K_2NiCa(NO_2)_6$ and subsequently weighed out that mass of $CaCl_2 \cdot 2H_2O$. Did Robin use too little or too much of the calcium salt, or did Robin's error have no effect on the mole ratio of K^+, Ni^{2+}, and Ca^{2+} in the reaction mixture? Explain your answer briefly.

3. Calculate the theoretical yield of $K_2NiCa(NO_2)_6$ from the reaction of 4.2794 g KNO_2, 1.1875 g of $NiCl_2 \cdot 6H_2O$, and 0.5824 g of $CaCl_2 \cdot 2H_2O$.

name ______________________ section ______________________ date __________

Summary Report on Experiment 11

Assigned transition metal ion (M') ______________________

Assigned nontransition metal ion (M) ______________________

Formula of reagent containing M' ______________________

Mass of container + reagent ______________________

Mass of container ______________________

Mass of reagent ______________________

Formula of reagent containing M ______________________

Mass of container + reagent ______________________

Mass of container ______________________

Mass of reagent ______________________

Mass of product and bottle ______________________

Mass of bottle ______________________

Mass of product ______________________

Theoretical yield ______________________

Percent yield ______________________

EXP
12

Determination of the Composition of Cobalt Oxalate Hydrate

Laboratory Time Required Three hours.

Special Equipment and Supplies

Analytical balance
Buret
Buret clamp
Litmus paper
0.02 M potassium permanganate, $KMnO_4$
0.1 M sodium oxalate, $Na_2C_2O_4$, standard solution
6 M sulfuric acid, H_2SO_4

Objective

In performing this experiment, students will use gravimetric analysis to determine the percentage of cobalt in cobalt oxalate hydrate. They will also do a redox titration to determine the percentage of oxalate in the compound.

Safety

Mouth pipetting is forbidden! Use the pipet bulb.

Sodium oxalate is a **poison.**
Potassium permanganate is **toxic.**

Potassium permanganate is a strong oxidizing agent.
Sulfuric acid is **corrosive** and may cause chemical burns.

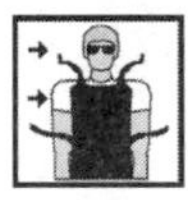

Because of the hazardous nature of the chemicals used, safety goggles are **mandatory**, and laboratory aprons are strongly advised.

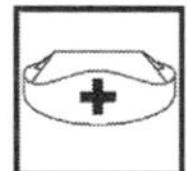

First Aid

If sodium oxalate or sulfuric acid gets on your skin, flush with water; then flush with aqueous sodium bicarbonate.

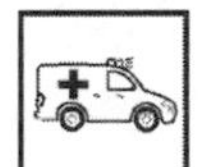

If you ingest sodium oxalate, drink a large quantity of water, followed by milk or milk of magnesia. Then see a doctor.

Use the eyewash fountain promptly and see a doctor if sulfuric acid gets into your eyes.

Aqueous $KMnO_4$ may leave brown stains on skin, but normally causes no serious health problems. Rinsing with water should be sufficient treatment for $KMnO_4$-stained skin. This may be followed by treatment with aqueous $NaHSO_3$, which will reduce the $KMnO_4$ or MnO_2 to the colorless Mn^{2+}.

Preamble

In this experiment, you will analyze a substance that has been given the tentative formula $Co_a(C_2O_4)_b \cdot cH_2O$. You may have synthesized the compound yourself in Experiment 10, or your instructor may provide you with a sample for analysis.

You will determine the percent cobalt (by mass) by a gravimetric method, and the percent oxalate (by mass) by an oxidation-reduction titration. When these percentages have been calculated, you will determine the percent water by difference (by subtracting the percentages of oxalate and cobalt from 100%). You will then use the composition by mass to calculate the actual formula of the compound. The experiment will be a test both of your proficiency and the applicability of the Law of Constant Composition. If the latter applies, the cobalt ions and oxalate ions combine in a definite ratio by mass. Because each ion has a characteristic mass, a definite number of cobalt ions combine with a definite number of oxalate ions. Hence *a* and *b* in the tentative formula should be integers.

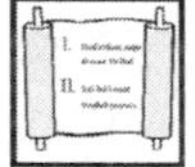

Principles

Thermal Decomposition of Cobalt Oxalate Hydrate

Because hydrates commonly lose all or part of their water when heated sufficiently, it would appear that the resulting mass loss could be used to determine the water content of hydrates that dehydrate easily. Unfortunately, this method cannot be used when thermal decomposition results in the formation of volatile products other than water. Metal oxalates, for example, generally decompose on being heated to yield gaseous CO or CO_2, as well as the metal or metal oxides. This is the case with cobalt oxalate hydrate, which decomposes to Co_3O_4 when heated in a crucible over a burner. Because the formula of the oxide is known, the mass of cobalt contained in a given mass of Co_3O_4 can be determined easily. (See Equation 12.1.) You may then use the mass of cobalt and the mass of the original sample to calculate the percent Co in the cobalt oxalate hydrate.

$$\#\text{g } Co_3O_4 \times \frac{1 \text{ mol } Co_3O_4}{240.80 \text{ g } Co_3O_4} \times \frac{3 \text{ mol Co}}{\text{mol } Co_3O_4} \times \frac{59.933 \text{ g Co}}{\text{mol Co}} = \#\text{g Co} \quad (12.1)$$

Determination of Oxalate

The reaction used in the volumetric determination of oxalate is shown in Equation 12.2.

$$5C_2O_4^{2-} + 2MnO_4^- + 16H^+ \rightarrow 10CO_2 + 2Mn^{2+} + 8H_2O \quad (12.2)$$

No indicator is needed because the presence of excess permanganate, which has an intense violet color, is easily seen. Thus, at the end point, a color change of the solution from colorless to pink or violet is normally observed. However, the presence of Co^{2+} ions, which have an orange-pink color in solution, makes the detection of the end point slightly more difficult. In such cases, the end point will be indicated by a color change from orange-pink to rose-pink or violet. This color change is more easily detected visually than it would appear from its description.

The reaction between permanganate ions and oxalate ions tends to be very slow, particularly at first and, for this reason, the oxalate solution is heated to about 60°C and kept at this temperature during the titration. The solution must not be boiled because the oxalate might decompose at higher temperatures. The first addition of permanganate will give the solution a violet color that will persist for some time because the reaction with oxalate ions is slow even at 60°C.

BE PATIENT. WAIT FOR EACH SMALL INCREMENT OF PERMANGANATE TO LOSE ITS COLOR BEFORE ADDING THE NEXT INCREMENT, OR YOU RISK PRECIPITATING MANGANESE DIOXIDE.

As the titration proceeds, the permanganate additions will be decolorized more and more rapidly. This happens because the Mn^{2+} ions that form during the reaction serve as a catalyst for the reaction. At the end point of the titration, the addition of one drop of permanganate solution will give the titration mixture a permanent color change from orange-pink to violet because there will be no oxalate left to reduce the permanganate ion.

Note that the reaction shown in Equation 12.2 does not occur in a 1:1 stoichiometric ratio. Therefore, the simple relation, $MV = MV$, cannot be used to find the molarity of the permanganate solution in the standardization trials or to find the chemical amount of oxalate contained in the titration of the cobalt oxalate sample. The appropriate conversions are illustrated in Equations 12.3 and 12.4

Standardization

$$M_{Na_2C_2O_4} V_{Na_2C_2O_4} \times \frac{2 \text{ mols } MnO_4^-}{5 \text{ mols } C_2O_4^{2-}} = \# \text{ mols } MnO_4^- \text{ titrated}$$

$$M_{MnO_4^-} = \frac{\# \text{ mols } MnO_4^- \text{ titrated}}{V_{MnO_4^-}} \qquad (12.3)$$

Analysis

$$M_{MnO_4^-} V_{MnO_4^-} \times \frac{5 \text{ mols } C_2O_4^{2-}}{2 \text{ mols } MnO_4^-} = \# \text{ mols } C_2O_4^{2-} \text{ in the sample} \qquad (12.4)$$

Determination of Water

As noted above, heating cobalt oxalate hydrate does more than simply drive off the water of hydration. However, the percent of water in the compound can be calculated by difference; that is, by subtracting the percentages of cobalt and oxalate from 100%.

Procedure

Determination of Cobalt

Powder the solid cobalt oxalate hydrate (either your dried precipitate or the sample provided by your instructor) and use the analytical balance to weigh (to four significant figures) a 0.3 g sample of the compound. Transfer this sample completely to a preweighed crucible. Heat the crucible and sample to red heat until the sample has decomposed to Co_3O_4, which is a stable, black solid. Allow the crucible to cool and determine the mass of the crucible plus residue. Calculate the mass percent cobalt. Repeat the determination. Report both answers and their average. To conserve time, you should begin the oxalate analysis while waiting for the crucible to cool.

Procedure in a Nutshell

Heat a known mass of cobalt oxalate hydrate in a preweighed crucible to convert the salt to Co_3O_4, a black solid. Standardize 0.02 M potassium permanganate by titrating a known volume of a standard solution of 0.1 M sodium oxalate that has been acidified. Use the standardized permanganate solution to titrate a known mass of cobalt oxalate hydrate.

Determination of Oxalate

Prepare 300 mL of 0.5 M H_2SO_4 by **carefully** adding, while stirring, 25 mL of 6 M H_2SO_4 to a 500-mL flask containing 275 mL of distilled water. Note the safety and first-aid procedures given at the beginning of this experiment.

Rinse a clean buret with two small portions of the approximately 0.02 M $KMnO_4$ solution and then fill the buret with the same solution. Accurately measure 5 to 10 mL of standard 0.1 M sodium oxalate solution into a 250-mL Erlenmeyer flask, add 50 mL of 0.5 M H_2SO_4 solution, and heat to about 60°C. Titrate with the potassium permanganate solution until a faint pink color, lasting about 30 seconds, is obtained. Calculate the molarity of the $KMnO_4$ solution. Repeat the standardization and calculate the average.

Weigh about 0.15 g of the cobalt oxalate hydrate to the nearest 0.1 mg and place it in a 250-mL Erlenmeyer flask. Add 50 mL of 0.5 M H_2SO_4 solution, stir to dissolve the solid, and heat to about 60°C. Titrate with the potassium permanganate solution until the color change signifying the end point is obtained. Calculate the mass percent oxalate in the compound. Repeat the titration and calculate the average.

When all permanganate titrations have been completed, rinse the buret with small portions of aqueous sodium sulfite. Be sure to remove all of the permanganate residue from the buret. Any purple solution left in the buret will eventually be converted to solid MnO_2 that can easily plug up the stopcock bore of the tip of the buret.

Disposal of Reagents

Add 3 M NaOH, by drops, to the titration solution containing Co^{2+} and Mn^{2+} until the solution tests basic to litmus. If no precipitate forms, dilute and discard the solution. If a precipitate forms, allow it to settle. Then decant the clear solution into another beaker for dilution and disposal. Pour the $Co(OH)_2$, $Mn(OH)_2$ slurry into the designated collection bottle. At a later time, this may be filtered, ignited, and bottled for disposal in a hazardous waste landfill site. The ignited product should be labeled Co_3O_4, Mn_3O_4.

Excess $KMnO_4$ solution should be reduced to Mn^{2+} by adding solid $NaHSO_3$ in small portions until the violet color disappears. The Mn^{2+} may then be precipitated as the hydroxide from a solution that is just basic and placed in a second collection bottle. The ignited product should be labeled Mn_3O_4.

Leftover sodium oxalate may be decomposed by acidification with H_2SO_4 and subsequent oxidation of oxalate to CO_2 with $KMnO_4$, following the procedure given in the experiment for the determination of oxalate. The resulting solution should be made just basic to precipitate $Mn(OH)_2$, which should then be placed in the appropriate collection bottle. The solution may be flushed down the drain. Note that you should coordinate your disposal of excess oxalate and permanganate so that one is used in the treatment of the other.

Any unused cobalt oxalate hydrate should be placed in a third collection bottle. This may be ignited when convenient and added to the Co_3O_4 bottle for disposal in a hazardous waste landfill.

Excess H_2SO_4 should be neutralized, diluted, and poured down the drain.

Questions

1. Using the experimental values for the percent cobalt, oxalate, and water in your compound, calculate the formula of the compound.

2. Write an equation for the thermal decomposition of cobalt oxalate hydrate.

name | section | date

Pre-Lab Exercises for Experiment 12

These exercises are to be completed after you have read the experiment but before you come to the laboratory to perform it.

1. Why is it unwise to boil the oxalate solution in preparation for its titration with permanganate?

2. A coordination compound is found by analysis to contain 21.79% Co, 18.90% NH_3, and 39.32% Cl. Assume the remainder, if any, is H_2O. Find the empirical formula for this coordination complex.

3. A 0.3288-g sample of a hydrocarbon (a compound of carbon and hydrogen) was burned in air. The sample was completely converted to 1.0562 g of carbon dioxide and 0.5046 g of water. Find the empirical formula of the compound.

4. In obtaining the answer to question 3, it is not necessary to find the percentages of carbon and hydrogen in the compound before obtaining the empirical formula. Why is it necessary to work with percentages in determining the empirical formula of cobalt oxalate hydrate as done in this experiment?

name	section	date

Summary Report on Experiment 12

Determination of Cobalt

	Trial 1	*Trial 2*
Mass of crucible and cobalt oxalate		
Mass of crucible		
Mass of cobalt oxalate		
Mass of crucible and residue		
Mass of crucible		
Mass of residue		
% Co in cobalt oxalate		
Average % Co		

Standardization of $KMnO_4$ *Solution*

Concentration of standard $Na_2C_2O_4$ solution		
Final reading of $Na_2C_2O_4$ buret		
Initial reading of $Na_2C_2O_4$ buret		
Volume of $Na_2C_2O_4$ solution dispensed		
Final reading of $KMnO_4$ buret		
Initial reading of $KMnO_4$ buret		
Volume of $KMnO_4$ solution required		
Molarity of $KMnO_4$ solution		
Average molarity of $KMnO_4$ solution		

Determination of Oxalate

Mass of container and cobalt oxalate	____________	____________
Mass of container	____________	____________
Mass of cobalt oxalate	____________	____________
Final reading of $KMnO_4$ buret	____________	____________
Initial reading of $KMnO_4$ buret	____________	____________
Volume of $KMnO_4$ solution used	____________	____________
% oxalate in cobalt oxalate	____________	____________
Average % oxalate	________________	

Determination of Water

Calculated % H_2O	____________
Empirical formula of cobalt oxalate hydrate	____________

EXP
13

The Gasimetric Analysis of a Nitrite Complex

Laboratory Time Required One and one-half hours.

Special Equipment and Supplies

Analytical balance
Aspirator
Small glass vials
Eudiometer tube
Glass beads
Sulfamic acid, NH_2SO_3H
$K_2MM'(NO_2)_6$

Objective In performing this experiment, the student will use gasimetric analysis to determine the % N in a nitrite compound.

Safety

Sulfamic acid is a skin and eye irritant.
The salts of Co^{2+}, Ni^{2+}, and Sr^{2+} are toxic.

Because of the toxicity of the salts and possible irritation to the eyes, protective eyewear is required; laboratory aprons are recommended.

First Aid

After exposure to any of these chemicals, thoroughly flush your skin or eyes with water. Remove chemicals from the stomach by drinking large amounts of water and inducing vomiting.

Preamble

In this experiment, you will perform a partial analysis of one of a group of compounds with the general formula $K_2MM'(NO_2)_6$, where M is an alkaline earth metal ion and M' is a divalent transition metal ion. You may have synthesized the compound you will analyze or your instructor may provide you with a sample for analysis. You will analyze the sample by decomposing it in a reaction with sulfamic acid and collecting the nitrogen gas evolved. You will use your data to calculate the percent nitrogen in your sample and compare your experimental value to the value calculated from the formula of your compound. This will permit you to determine whether your compound's formula agrees with the general formula cited above.

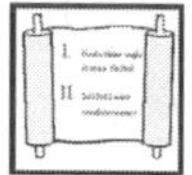

Principles

Nitrite Analysis

In the presence of acids, the nitrite complexes $K_2MM'(NO_2)_6$ readily release their nitrite ligands to form HNO_2 as shown in Equation 13.1.

$$M(NO_2)_6^{4-} + 6H^+ \rightarrow 6HNO_2 + M^{2+} \tag{13.1}$$

The HNO_2 thus formed reacts quantitatively with sulfamic acid according to the reaction shown in Equation 13.2.

$$HNO_2 + NH_2SO_3^- \rightarrow N_2 + HSO_3^- + H_2O \tag{13.2}$$

After you have collected the nitrogen gas and measured its volume, you can use the gas laws to calculate the number of moles of nitrite present in the sample. The apparatus needed is shown in Figure 13.1.

As shown in Figure 13.1, the nitrogen is collected by being bubbled through water into a eudiometer tube. As a result, the gas collected is wet and its total pressure is the sum of the pressure of the dry gas and the water vapor pressure. The vapor pressure of water depends, in turn, on the temperature. A table of vapor pressures of water at various temperatures is given in Appendix B.

In most cases, the volume of gas collected will not be sufficient to lower the water level inside the eudiometer tube to that of the water in the beaker in which the tube is mounted. Thus, the barometric pressure will not be completely balanced by the pressure of the wet gas and will still be able to support a column of water. The height of this column is designated as Δh in Figure 13.1.

The relation between the pressure of the dry gas, barometric pressure, water vapor pressure, and the height of the water column is given in Equation 13.3. The factor 13.6 is introduced to convert the height of the water column to the equivalent height of a mercury column. This is necessary because the other pressures are measured in units of torr (1 torr = 1 mm Hg).

$$P_{N_2} = P_{atm} - P_{H_2O} - \frac{\Delta h}{13.6} \tag{13.3}$$

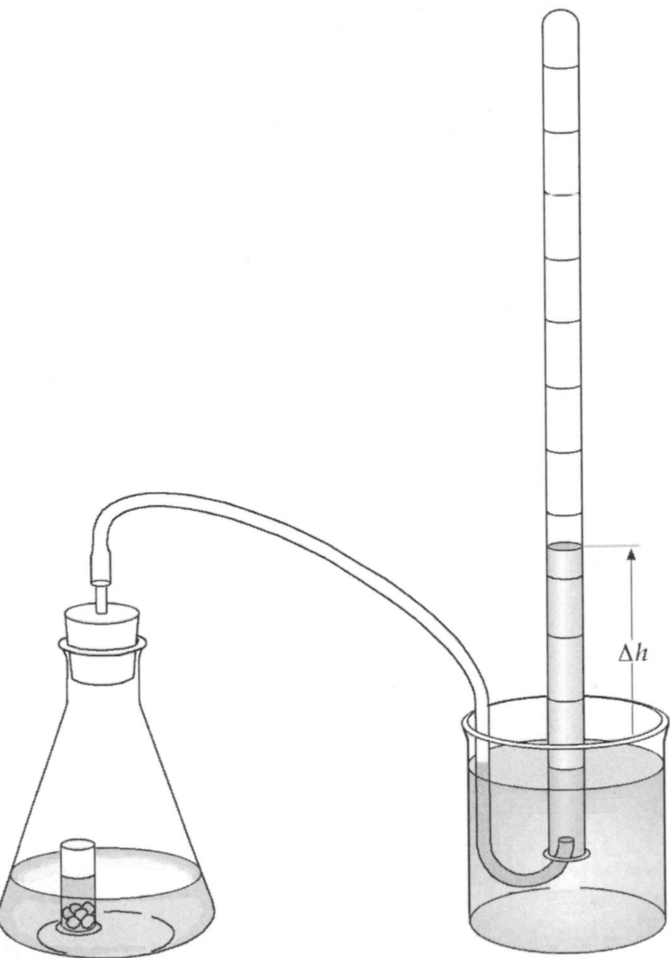

Figure 13.1 Reaction flask with eudiometer tube

Procedure

Procedure in a Nutshell

Place a weighed amount of nitrite complex into a small vial. Lower the vial into a flask containing an aqueous solution of sulfamic acid, making sure that the vial remains upright. Connect the flask to a water-filled eudiometer tube. Allow the sulfamic acid to mix with the contents of the vial so that the nitrogen produced is collected in the eudiometer tube.

If you did not perform Experiment 11, obtain a sample of a compound salt from your instructor, who will give you its formula. Calculate the mass of compound that should be used in order to produce 30 to 40 mL of nitrogen gas and accurately weigh that amount into each of two glass vials. The vials should be small enough so that they can be upset easily after being inserted upright into a 50-mL Erlenmeyer flask. Dissolve 0.5 g of sulfamic acid in a 50-mL Erlenmeyer flask, using 20 mL of water. Carefully place one vial upright in the flask without mixing the solid and solution, and stopper the flask with a one-hole rubber stopper containing a short length of glass tubing. (Distilled water or glass beads may be added to the vial to reduce its tendency to tip over prematurely.) Attach one end of a rubber tube to the glass tube in the one-hole rubber stopper and insert the other end into a water-filled eudiometer tube. Mix the solid and solution by tilting the Erlenmeyer flask and then shaking the mixture gently until the solid has all reacted. Wait approximately 12 minutes and then read and record the volume of gas collected in the eudiometer tube and also the difference (in millimeters) between the water level inside the eudiometer tube and the water level outside the tube. Use a meter stick to make the latter measurement. Repeat the analysis using the solid contained in the second vial.

Compute the mass percent nitrogen for each sample and average the results. Show your calculations clearly in your report, including corrections for the presence of water vapor in the gas collected and for hydrostatic pressure caused by the difference in water levels. Calculate the theoretical percent nitrogen and your percent error and include these in your report.

Disposal of Reagents

The small amounts of hazardous chemicals remaining in solution make it impractical to isolate the ions as insoluble salts. Consequently, the recommended disposal procedure for all solutions used in this experiment is to dilute them greatly and flush them down the drain with water. Excess nitrite salts can be stored in desiccators for analysis by other groups of students.

name section date

Pre-Lab Exercises for Experiment 13

These exercises are to be completed after you have read the experiment but before you come to the laboratory to perform it.

As a result of the reaction between a nitrite compound and sulfamic acid, a 43.7 mL sample of nitrogen was collected over water at 24.0°C. The height of the water column inside the eudiometer tube was 126.9 mm higher than the surface of the water in the surrounding beaker. Barometric pressure was 739.2 torr.

1. Show your work in finding the partial pressure of the nitrogen gas, P_{N_2} .

2. Show your work in finding the mass of the nitrogen gas collected.

3. The original sample, which decomposed to produce the nitrogen gas, weighed 0.1198 g. Find the percent nitrogen in the sample.

name ______ section ______ date ______

Summary Report on Experiment 13

	Trial 1	Trial 2	Trial 3
Mass of bottle (after removal of sample)	______	______	______
Mass of bottle (before removal of sample)	______	______	______
Mass of sample in vial	______	______	______
Volume of gas collected	______	______	______
Δh	______	______	______
Barometric pressure	______	______	______
Room temperature	______	______	______
Vapor pressure of water	______	______	______
Moles of nitrogen collected	______	______	______
Mass of nitrogen in sample	______	______	______
Percent nitrogen in sample	______	______	______
Average percent nitrogen		______	
Theoretical percent nitrogen		______	
Percent error		______	

EXP
14

The Burning of a Candle

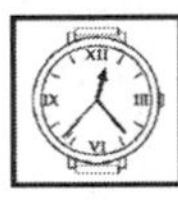

Laboratory Time Required

Two hours.

Special Equipment and Supplies

Analytical balance
Top-loading balance
Glass plates
Shallow pan
Matches
Ice
Glass-marking pen
Watch glass
Timer
Candle, 0.8 cm × 10 cm
Bromothymol blue indicator
Limewater, saturated $Ca(OH)_2(aq)$

Objective

In performing this experiment, students will consider some of the complexities involved in analyzing the "simple" process of burning a candle.

Safety

Exercise due caution in working with flames. Do not leave flames unattended. Restrain long sleeves and hair near flames.

Limewater is a moderately strong base. Wear safety glasses when working with limewater.

Rinse hands thoroughly with water after contact with limewater.

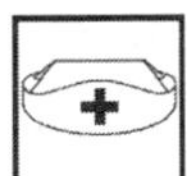

First Aid

Minor burns may be treated by immersing the area in cool water.

More extensive burns will require a doctor's care.

If limewater gets in the eyes, flush them with water for at least 20 minutes and then seek medical attention.

Limewater splashed on the skin should be washed off promptly.

Preamble

The study of combustion reactions was of enormous importance in the development of modern chemistry. This experiment on the burning of a candle is a deceptively simple one that will allow you to develop your powers of observation and challenge you to analyze your data logically.

Principles

Although electric lights and central heating have decreased the number of uses for fire in our homes, we still use combustion every day of our lives. We drive in cars powered by internal combustion engines; we may burn gas to cook our food or heat our homes; and we may light candles for festive occasions, religious ceremonies, or romantic evenings. If fire is important to us, it was all the more so to the ancients. The Greeks believed that fire was stolen from the gods by Prometheus, who was made to bear a terrible punishment for his crime.

Despite the fact that people have used fire for thousands of years, the true nature of combustion reactions was shrouded in mystery until the latter part of the eighteenth century. Lavoisier's discovery that combustion consists of the uniting of oxygen with other substances ranks with Dalton's atomic theory as the foundation of modern chemistry.

In this experiment, you will investigate the burning of a candle, collecting evidence to support or refute Lavoisier's theory of combustion. Qualitative observations will test the hypothesis that, during combustion, the carbon and hydrogen present in candle wax combine with oxygen of the air to form water, carbon dioxide, and possibly some elemental carbon as well. The physical properties of these reaction products will be used for identification. You might wish to review Experiment 7, if necessary, to recall the reactions of aqueous CO_2, with acid-base indicators and with limewater. You will use bromothymol blue indicator, which is blue in basic solutions and yellow in acidic solutions.

You will also study the reaction quantitatively to determine the chemical amount (in moles) of candle wax (assumed to be $C_{21}H_{44}$) and of oxygen consumed when the candle is burned in a limited amount of air. The experimental mole ratio will be compared with that obtained from a balanced equation for the reaction, which you will be asked to write.

The quantity of wax consumed will be determined by weighing the candle before and after the reaction. It will be slightly more difficult to find the quantity of oxygen consumed, but you can calculate this quantity using the Ideal Gas Law and Dalton's Law of Partial Pressures, shown in Equations 14.1 and 14.2, respectively.

$$PV = nRT \qquad (14.1)$$

$$P_{tot} = P_A + P_B \qquad (14.2)$$

You will begin the reaction by placing a flask of known volume over a burning candle that is standing in a dish of water. The number of moles of air in the flask *(n)* can be calculated using Equation 14.1, by substituting the volume of the flask in liters *(V)*, the air pressure in atmospheres *(P)*, the Kelvin temperature *(T)*, and the value of the universal gas constant ($R = 0.0806\ \mathrm{L \cdot atm/mol \cdot K}$). The initial air pressure in the flask will be equal to atmospheric pressure in the laboratory, which is measured using a barometer.

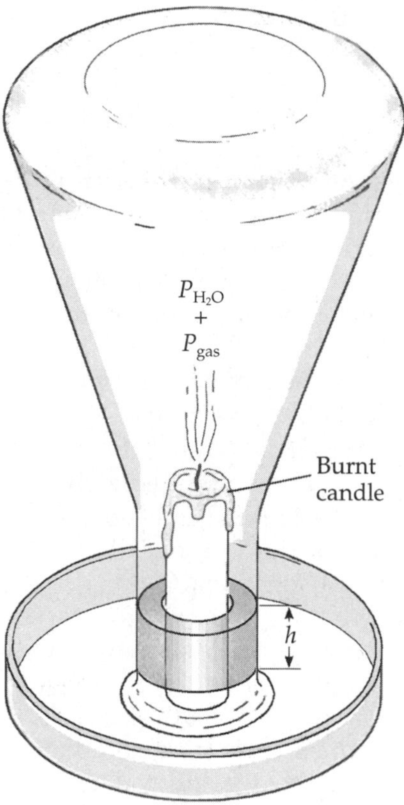

Figure 14.1 The result of burning a candle in a confined volume

As the candle burns, consuming oxygen, the water level in the flask will rise (Figure 14.1). This indicates that the pressure exerted by the gases in the flask has decreased to a value below atmospheric pressure. The water will rise until the pressure in the flask (given in Equation 14.3) equals atmospheric pressure.

$$P_{atm} = P_{\text{in flask}} = P_{air} + P_{H_2O} + \frac{\Delta h}{13.6} \tag{14. 3}$$

In Equation 14.3, all pressures are given in units of torr, or mm Hg (760 torr = 1 atm). The symbol, P_{H_2O}, represents the partial pressure of water in the gas or vapor phase, resulting from the evaporation of liquid water. Fortunately, P_{H_2O} is constant at a given temperature. You may therefore obtain the value of P_{H_2O} from the table in Appendix B.

The final term in Equation 14.3, $\Delta h/13.6$, represents the pressure caused by the internal water level being Δh mm higher than the water level outside the flask. The 13.6 divisor is needed to convert from mm of H_2O to mm of Hg.

The number of moles of air in the flask after the candle is burned can be calculated using Equation 14.1, substituting the appropriate values for the air pressure, volume, and temperature. The air pressure can be calculated using Equation 14.3, as previously discussed. You can determine the final air volume by marking the water level on the flask after burning the candle and then measuring the volume of water needed to fill the flask to that mark. As a first approximation, you may assume that the amount of oxygen consumed by the burning candle corresponds to the difference between the initial and final amounts of air in the flask.

Procedure

Procedure in a Nutshell

Burn a candle in open air and try to detect some of the products of combustion. Burn one (and then, two) candles under an Erlenmeyer flask and measure the time it takes for the single candle (and then, both candles) to be extinguished. Burn a single candle under an Erlenmeyer flask while the candle is standing in a pan of water. Determine the height of the water column that enters the flask. Determine the state of the carbon dioxide produced in combustion via reactions of indicators placed in the water in the pan. Try to relate the mass of the candle consumed in combustion with the amount of oxygen used and to the changes in pressure that occur in the flask in the course of the combustion reaction.

Obtain a candle and matches from your instructor. Hold the candle over a lit match for a few seconds so that the bottom of the candle begins to soften. Press the candle firmly onto a glass plate while the candle's bottom is still soft. The candle should be able to stand upright without support.

Light the candle. Hold a beaker in the flame for a few moments. Then allow the beaker to cool and wipe it with a white tissue. Note any evidence of the deposition of carbon on the beaker.

Fill a watch glass with ice and hold it for a few moments about 2 inches above the flame. Note any evidence for the condensation of water on the underside of the watch glass.

Extinguish the flame. Weigh the candle and plate on a balance. Light the candle once again and immediately invert a 250-mL Erlenmeyer flask carefully over the candle. The candle should continue to burn for a few seconds and then should go out. After the flame has burned out, weigh the candle and plate once again.

Fill the Erlenmeyer flask with water and weigh the water-filled flask on a top-loading balance. Assume that the density of water is 1.00 g/mL and calculate the volume of the flask. Determine the chemical amount of oxygen in the flask, assuming that air is 20% oxygen. Compare the amount of oxygen consumed (assuming complete reaction of $C_{21}H_{44}$ and oxygen gas to yield carbon dioxide and water).

Repeat the process of lighting the candle and inverting a (dry) flask over it; use the timer to determine how long it takes for the flame to go out. Put another candle on the plate, light them both, and invert a flask over them. Use the timer, as before, to determine how long it takes for the flame to be extinguished.

Place a single candle on a glass plate and put the assemblage into a metal pan. Add water to the pan until the glass plate is just barely submerged about one-quarter inch. Add some bromothymol blue indicator to the water and stir carefully so that the blue color of the indicator spreads throughout the water but the candle is not dislodged. Relight the candle and once again invert a dry 250-mL Erlenmeyer flask over it. Record your observations. After the candle has been extinguished, use a meter stick to measure the difference between the water levels inside and outside of the flask. Record this value of Δh . Use a glass-marking pen to place a line on the flask at the height of the water level inside the flask. Carefully lift the flask, with the glass plate remaining on its mouth, and invert and shake the flask gently. Record your observations.

Obtain another candle and repeat the procedure of mounting it on a glass plate and placing the plate in a pan. However, this time fill the pan with limewater until the plate is submerged about one-quarter inch. Light the candle and invert a dry, 250-mL Erlenmeyer flask over it. Record your observations. After the flame is extinguished, repeat the procedure of inverting and shaking the flask (with the glass plate over its mouth). Record your observations.

Finally, fill your marked 250-mL Erlenmeyer flask to the pen mark with water. Weigh the flask and water on a top-loading balance and calculate the volume of water that entered the flask. Be sure to record the values of the barometric (atmospheric) pressure and ambient (room) temperature. Answer the questions on the Summary Report Sheet.

Disposal of Reagents

Be sure all matches are extinguished before they are discarded in the wastebaskets. Save the candles for reuse. Dispose of the limewater by flushing it down the drain with large amounts of water. Water containing bromothymol blue may also be poured down the drain.

Questions

1. This experiment is based on a phenomenon that is very often demonstrated in elementary school science classes. The fact that the water rises is attributed to its "replacing the 20% of air that is oxygen." Would that explanation agree with your results? Explain why or why not.

2. Is the water produced in the burning of a candle in its gaseous or liquid phase? What evidence do you have for your answer?

3. a. Consider the case of the candle burning while its base is submerged in water. Suppose the carbon dioxide dissolves in the water as soon as it is produced. Further suppose that only liquid water is produced as the candle burns. How high should the column of water rise under these circumstances?

b. Suppose that the carbon dioxide does not dissolve, but remains in the gas phase. How high should the water column rise, if the water produced is actually in the liquid phase?

c. Suppose that both the carbon dioxide and the water produced in the combustion of the candle are in the gas phase. How high should the water column rise under these circumstances?

4. List at least five sources of error in this experiment that would affect your knowledge of the quantitative relationship between the amounts of wax and oxygen consumed in the combustion of the candle and the amounts of carbon dioxide and water produced.

name ______ section ______ date ______

Pre-Lab Exercises for Experiment 14

These exercises are to be completed after you have read the experiment but before you come to the laboratory to perform it.

1. Balance the equations shown below. Each shows the combustion of a hydrocarbon or carbohydrate.

$$CH_4 + O_2 \rightarrow CO_2 + H_2O$$

$$C_8H_{18} + O_2 \rightarrow CO_2 + H_2O$$

$$C_{12}H_{22}O_{11} + O_2 \rightarrow CO_2 + H_2O$$

2. Describe an experiment that would distinguish between a compound that is a hydrocarbon (a compound of C and H only) and a compound that is a carbohydrate (a compound of C, H, and O only).

3. A sample of nitrogen was collected over water on a day when the atmospheric pressure was 735.6 torr and the temperature was 22.0°C. A volume of 39.3 mL of gas was collected. The water level inside the collection vessel was 256 mm above the water level outside the vessel. Find the amount of N_2 (in moles) that was collected (1 atm = 760 torr).

name section date

Summary Report on Experiment 14

What did you observe when the beaker that had been held in the flame was wiped with a tissue?

What can you infer from this?

What did you observe when the ice-filled watch glass was held in the flame?

What can you infer from this?

Mass of candle/glass plate after burning

Mass of candle/glass plate before burning

Change in mass

What did you observe when the 250-mL Erlenmeyer flask was inverted over one candle in the absence of water?

What can you infer from this?

What did you observe when the 250-mL Erlenmeyer flask was inverted over two candles in the absence of water?

What can you infer from this?

What did you observe when the 250-mL Erlenmeyer flask was inverted over the candle in the presence of water containing bromothymol blue indicator?

Value of Δh

What did you observe when the flask containing water and bromothymol blue was shaken?

____

____

What can you infer from this? ______________________________

What did you observe when the flask was inverted over the candle in the presence of limewater?

____

____

What did you observe when the flask containing the limewater was shaken?

____

____

What can you infer from this? ______________________________

Volume of air before candle was burnt ______________

Volume of air after candle was burnt ______________

Atmospheric pressure ______________

Room temperature ______________

Write a balanced equation for the combustion of candle wax, $C_{21}H_{44}$.

How many moles of candle wax burned? ______________

From the equation, how many moles of oxygen were consumed in combustion of the candle wax? ______________

Using the Ideal Gas Law, calculate the number of moles of air originally present in the 250-mL flask. ______________

Calculate the number of moles of air present in the flask after the candle was burnt. ______________

Calculate the number of moles of oxygen consumed based on the Ideal Gas Law calculations. ______________

The Vapor Pressure of Water

EXP 15

Laboratory Time Required

Two hours.

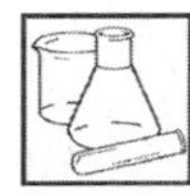

Special Equipment and Supplies

Thermometer
Hot plate or burner
Graduated cylinder, 10-mL
Tall-form beaker, 1000-mL
Ice

Objective

In performing this experiment, the student will explore the relationship between the vapor pressure of water and temperature, and evaluate ΔH°_{vap} for water.

Safety

This experiment involves moving a beaker full of hot water. Always remain alert and be cautious when handling hot water. Never leave a burner flame unattended.

Burns or electrical shock may be caused by electric heating devices that are poorly maintained or carelessly used. Avoid electric shock by using care when plugging in the power cord and by not using instruments that have frayed cords or that are wet from spilled liquids.

Because hot liquids pose a danger to your eyes, you should wear safety goggles.

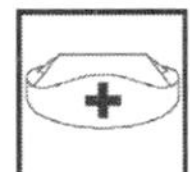

First Aid

You may soothe burnt fingers by immersing them in cool water.

Seek medical attention for serious burns.

In case of electric shock, separate the victim from the source of electricity by turning off the power at the switch box, removing the power cord with a nonconducting tool, or by other means.

If the victim is not breathing or has no heartbeat, CPR should be administered quickly by a qualified person. Get professional help *immediately*.

Preamble

Although phase changes are not chemical changes, the examination of phenomena such as vaporization is an important part of the study of chemistry. A liquid's volatility, enthalpy of vaporization, and normal boiling point are characteristics that reflect the intermolecular forces present in the liquid. This experiment employs simple apparatus in the study of a one-component system, water.

Principles

The atoms and molecules of any liquid are in constant motion, continually changing their molecular speeds and kinetic energies as a result of collisions. At any given temperature, a number of molecules may have sufficient kinetic energy to escape from the liquid at the surface, evaporating into the space above the liquid. Consequently, the particles remaining in the liquid have lower kinetic energy, and the temperature of the liquid decreases, unless the liquid absorbs energy from its surroundings. If the liquid is in an open container, allowed to absorb heat from the surroundings to maintain a constant temperature, evaporation will continue until no more liquid remains. If, however, the liquid evaporates in a closed container, an equilibrium is established in which the rate of escape from the liquid is balanced by the rate at which gas phase particles lose energy and return to the liquid phase. The pressure exerted on the walls of the container when equilibrium has been established is called the **equilibrium vapor pressure** of the liquid.

The value of the equilibrium vapor pressure increases with temperature for all liquids. When the vapor pressure reaches the value of the external pressure, the liquid boils. The temperature at which the vapor pressure equals 760 torr (one standard atmosphere) is called the normal boiling point of the liquid.

In this experiment, you will study the relationship between the vapor pressure of water and temperature by monitoring the volume of an air bubble that is surrounded by a water bath. At temperatures above $5°C$, water has an appreciable vapor pressure and Dalton's Law of Partial Pressures is used to relate the partial pressure of air, the vapor pressure of water, and atmospheric pressure. (See Equation 15.1.)

$$P_{atm} = P_{air} + P_{H_2O} \qquad (15.1)$$

At temperatures below $5°C$, the vapor pressure of water is negligibly small. Therefore, at low temperature, the bubble may be considered to contain only air. The Ideal Gas Law can be used to relate the amount of air (n_{air}) to the volume (V) of the bubble, the bath temperature (T), and the atmospheric pressure (P), as shown in Equation 15.2.

$$n_{air} = \frac{PV}{RT} \qquad T < 278\text{ K} \qquad (15.2)$$

At temperatures above $5°C$, the bubble becomes saturated with water vapor. However, the amount of air contained in the bubble is constant. The Ideal Gas Law can once again be used to obtain the partial pressure of air (P_{air}) from the number of moles (n_{air}) of air in the bubble, the volume (V) of the bubble, and the bath temperature (T), as shown in Equation 15.3.

$$P_{air} = \frac{n_{air}RT}{V} \qquad T > 278\text{ K} \qquad (15.3)$$

The value of P_{H_2O}, the vapor pressure of water, is then obtained from Equation 15.1. Once the values for the vapor pressure at different temperatures have been obtained, they can be used to find two characteristic properties of water—its normal boiling point and its enthalpy of vaporization ΔH_{vap}. The enthalpy of vaporization is the heat that must be supplied to evaporate a mole of water at constant pressure. The relationship of ΔH_{vap} to the vapor pressure at different temperatures is given in Equation 15.4, where P_1 and P_2 represent the vapor pressure of water at temperature T_1 and T_2, respectively. The symbol "ln" denotes the natural logarithm. The constant, R, is the ideal gas constant with the value of $8.314\ \text{joule/K}\cdot\text{mole}$ rather than the value $0.08206\ \text{L}\cdot\text{atm/K}\cdot\text{mole}$, which would be used in Equations 15.2 and 15.3.

$$\ln P_2 - \ln P_1 = \frac{-\Delta H_{vap}}{R}\left(\frac{1}{T_2} - \frac{1}{T_1}\right) \qquad (15.4)$$

The value of ΔH_{vap} is obtained by plotting $\ln P$ versus $1/T$. Such a plot should be a straight line, with slope equal to $-\Delta H_{vap}/R$. Once ΔH_{vap} has been obtained, one may solve the equation to find the value of T at which P would equal 760 torr.

Procedure

Procedure in a Nutshell

Study the variation of volume with temperature of an air-bubble trapped in a graduated cylinder inverted in a water-bath. Use the Ideal Gas Law to relate the volume/temperature data to the vapor pressure of water at each temperature studied.

Obtain a 10-mL graduated cylinder and a beaker large enough for the cylinder to be submerged in it. Fill the beaker half full with distilled water. Put enough distilled water in the 10-mL graduated cylinder to fill the cylinder to 90% capacity (ignoring graduations). Place your finger over the mouth of the graduated cylinder and invert the cylinder in the beaker. An air bubble, 4 to 5 mL in volume, should remain in the cylinder. Add distilled water until the graduated cylinder is covered completely, as shown in Figure 15.1. Heat the water in the beaker to 75° or 80°C. The air sample should be allowed to extend beyond the calibrated portion of the cylinder without escaping. Remove the beaker from the heat when the desired temperature has been reached. Start recording the volume of the bubble and the water temperature when the air sample is contained completely within the calibrated portion of the cylinder. Take readings every 3°C until the water temperature has cooled to 50°C. Then add ice to the beaker to lower the temperature below 5°C. Record the volume of the air bubble at low temperature. Also record the value of the barometric pressure.

Use the data obtained at low temperature to find the number of moles of air in your bubble. Then find the partial pressure of air at each of the higher temperatures for which you recorded volume data. Also find the vapor pressure of the water at each of those temperatures. Prepare a table with columns for t(°C), T(K), $1/T$, P_{H_2O}, and $\ln P$. Use your tabulated results to prepare a plot of $\ln P$ vs. $1/T$. Use your plot to find the value of ΔH_{vap} for water. Predict the normal boiling point of water.

Disposal of Reagents

The water and ice in this experiment can be discarded in the sink.

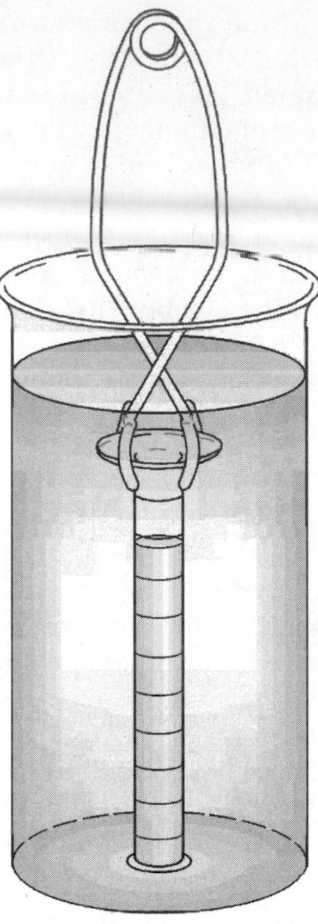

Figure 15.1 Apparatus used to determine the vapor pressure of water

Questions

1. What is the uncertainty associated with each of your volume observations? How does this affect your value of P_{H_2O} at 50°C ?

2. You could obtain ΔH_{vap} by inserting the data from two P_{H_2O} measurements into Equation 15.4, or from a plot of data from 8–10 measurements, as in this experiment. Which procedure is better? Why?

name　　　　section　　　　date

Pre-Lab Exercises for Experiment 15

These exercises are to be completed after you have read the experiment but before you come to the laboratory to perform it.

1. The vapor pressure of acetone at 39.5°C is 400 torr. At 7.7°C, the vapor pressure of acetone is 100 torr. Use these data and Equation 15.4 to find the value of ΔH_{vap} for acetone.

2. Kim Pupil performed this experiment and prepared a plot of $\ln P$ versus $1/T$ with a temperature scale ranging from $1/273$ K to $1/353$ K. Fran Teacher penalized Kim 10 points. What was wrong with Kim's graph?

3. List three assumptions made in obtaining results from the data collected in the course of performing this experiment.

name	section	date

Summary Report on Experiment 15

Observations

Barometric Pressure ______________ torr

t (°C)	V (mL)

Tabulated Results

n_{air} ______________

t (°C)	T (K)	$1/T$ (K^{-1})	P_{air} (torr)	P_{H_2O} (torr)	$\ln P_{H_2O}$

ΔH_{vap}, J/mol ______________

Predicted normal boiling point of water, °C ______________

EXP
15P

Vapor Pressure of Liquids[1]

Preamble

In this experiment, you will investigate the relationship between the vapor pressure of a liquid and its temperature. When a liquid is added to the Erlenmeyer flask shown in Figure 15P.1, it will evaporate into the air above it in the flask. Eventually, equilibrium is reached between the rate of evaporation and the rate of condensation. At this point, the vapor pressure of the liquid is equal to the partial pressure of its vapor in the flask. Pressure and temperature data will be collected using a pressure sensor and a temperature probe. The flask will be placed in water baths of different temperatures to determine the effect of temperature on vapor pressure. You will also compare the vapor pressure of two different liquids, ethanol and methanol, at the same temperature.

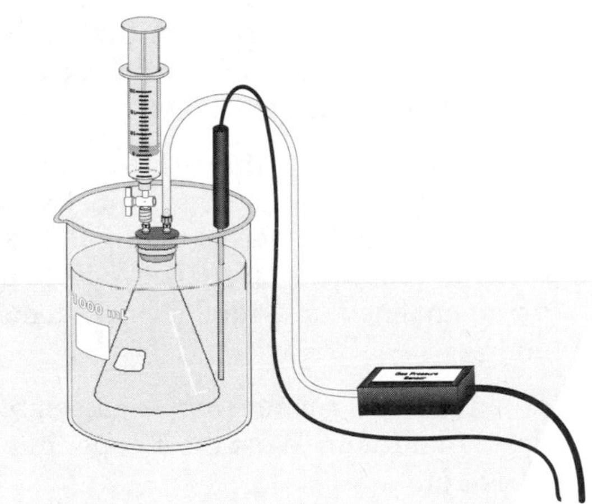

Figure 15P.1

Materials

CBL 2 interface
TI Graphing Calculator
Vernier gas pressure sensor or pressure sensor
Vernier temperature probe
Four 1-liter beakers
20-mL syringe
Rubber-stopper assembly
Plastic tubing with two connectors
Two 125-mL Erlenmeyer flasks
Methanol
Ethanol
Ice

[1] Adapted from Holmquist, D.D., J. Randall, D.L. Votz, *Chemistry with Calculators: Chemistry Experiments Using Vernier Sensors with Texas Instruments CBL 2™* (Vernier Software and Technology, Beaverton, Oregon: 2000). Used with permission.

Procedure

1. Obtain and wear goggles! **CAUTION:** The alcohols used in this experiment are flammable and poisonous. Avoid inhaling their vapors. Avoid contacting them with your skin or clothing. Be sure there are no open flames in the lab during this experiment. Notify your TA immediately if an accident occurs.

2. Use 1-liter beakers to prepare four water baths, one in each of the following temperature ranges: 0 to 5°C, 10 to 15°C, 20 to 25°C (use room temperature water), and 30 to 35°C. For each water bath, mix varying amounts of warm water, cool water, and ice to obtain a volume of 800 mL in a 1-L beaker. To save time and beakers, several lab groups can use the same set of water baths.

3. Prepare the temperature probe and pressure sensor for data collection.

a. Plug the temperature probe into Channel 1 of the CBL 2 interface.

b. Plug the pressure sensor into Channel 2 of the interface.

c. Identify which type of Vernier pressure sensor you are using:

- Newer Vernier gas pressure sensors have a white stem protruding from the end of the sensor box.
- Older Vernier pressure sensors have a 3-way valve at the end of a plastic tube leading from the sensor box. Before proceeding with the next step, align the blue handle with the stem of the 3-way valve that will *not* be used, as shown in Figure 15P.2—this will close this stem.

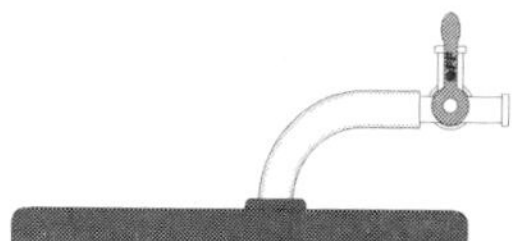

Figure 15P.2

d. Obtain a rubber-stopper assembly with a piece of heavy-wall plastic tubing connected to one of its two valves. Attach the connector at the free end of the plastic tubing to the open stem of the pressure sensor with a clockwise turn. Leave its two-way valve on the rubber stopper open (lined up with the valve stem as shown in Figure 15P.3) until Step 9.

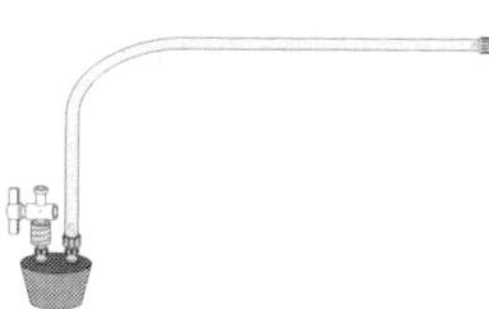

Figure 15P.3

e. Insert the rubber-stopper assembly into a 125-mL Erlenmeyer flask. **Important:** Twist the stopper into the neck of the flask to ensure a tight fit.

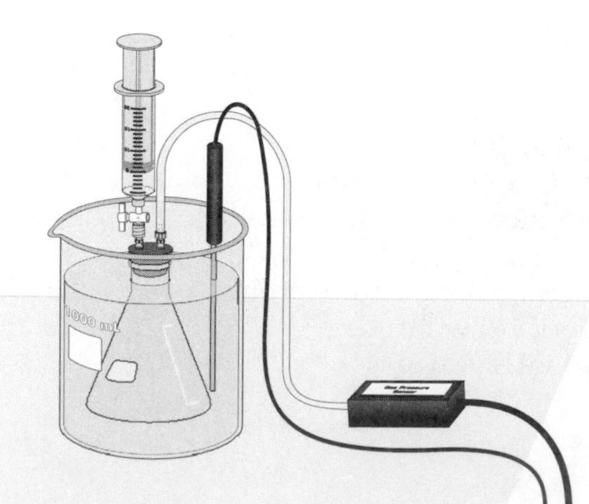

Figure 15P.4

4. Turn on the calculator and start the DataMate program. Press CLEAR to reset the program.

5. Set up the calculator and interface for a temperature probe in CH 1 and a pressure sensor in CH2.

a. Select SETUP from the main screen.

b. If the calculator displays a temperature probe in CH 1 and a pressure sensor set to kPa in CH2, proceed directly to Step 6. If it does not, continue with this step to set up your sensor manually.

c. Press ENTER to select CH 1.

d. Select TEMPERATURE from the SELECT SENSOR menu.

e. Select the temperature probe you are using (in °C) from the TEMPERATURE menu.

f. Press ▼ once, then press ENTER to select CH2.

g. Select PRESSURE from the SELECT SENSOR menu.

h. Select the correct pressure sensor (GAS PRESSURE SENSOR or PRESSURE SENSOR) from the PRESSURE menu.

i. Select the calibration listing for units of kPa.

6. Set up the data-collection mode.

a. Advance the cursor to MODE using the ▲ key, then press ENTER to select it.

b. Select SELECTED EVENTS from the SELECT MODE menu.

c. Select OK to return to the main screen.

7. The temperature and pressure readings should now be displayed on the calculator screen. While the two-way valve above the rubber stopper is still open, record the value for atmospheric pressure in your data table (round to the nearest 0.1 kPa).

8. Finish setting up the apparatus shown in Figure 15P.4:

a. Obtain a room-temperature water bath (20-25°C).

b. Place the temperature probe in the water bath.

c. Hold the flask in the water bath, with the entire flask covered as shown in Figure 15P.4.

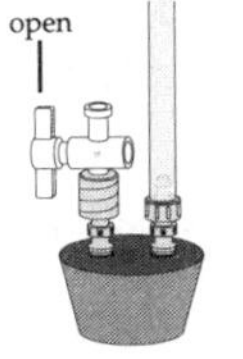

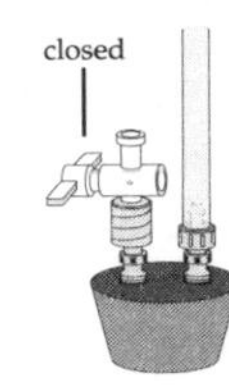

Figure 15P.5

d. After 30 seconds, close the 2-way valve *above the rubber stopper* as shown in Figure 15P.5—do this by turning the white valve handle so it is perpendicular with the valve stem itself.

9. Obtain the methanol container and the syringe. Draw 3 mL of the methanol up into the syringe. With the two-way valve still closed, screw the syringe onto the two-way valve, as shown in Figure 15P.4.

10. Introduce the methanol into the Erlenmeyer flask.

a. Open the 2-way valve above the rubber stopper—do this by turning the white valve handle so it is aligned with the valve stem. (See Figure 15P.5.)

b. Squirt the methanol into the flask by pushing in the plunger of the syringe.

c. *Quickly* return the plunger of the syringe back to the 3-mL mark of the syringe, then close the 2-way valve by turning the white valve handle so it is perpendicular with the valve stem.

d. Remove the syringe from the 2-way valve with a counter-clockwise turn.

11. To monitor and collect temperature and pressure data:

a. Select START to begin data collection.

b. The flask should still be submerged in the water bath.

c. When the temperature and pressure readings displayed on the calculator screen have both stabilized, equilibrium between methanol liquid and vapor has been established. Select ENTER on the calculator to store the first temperature-pressure data pair.

12. To collect another data pair using the 30–35°C water bath:

a. Place the Erlenmeyer flask assembly and the temperature probe into the 30–35°C water bath. Make sure the entire flask is covered.

b. When the temperature and pressure readings displayed on the calculator screen have both stabilized, select ENTER on the calculator to store the second data pair.

13. For Trial 3, repeat the Step-12 procedure, using the 10–15°C water bath. Then repeat the Step-12 procedure for Trial 4, using the 0–5°C water bath. After you have collected the fourth and last data pair, press [STO▶] to stop data collection. Remove the flask and the temperature probe from the last water bath.

14. Open the side valve of the pressure sensor so the Erlenmeyer flask is open to the atmosphere. Remove the stopper assembly from the flask and dispose of the methanol as directed by your teacher.

15. Press [▼] twice, then press [ENTER] to choose CH2 VS. CH1. Examine the data points along the displayed graph of pressure *vs.* temperature (°C). As you move the cursor right or left, the temperature (X) and pressure (Y) values of each data point are displayed below the graph. Record the data pairs in your data table. Round the pressure to the nearest 0.1 kPa and the temperature to the nearest 0.1°C.

16. Obtain another clean, dry 125-mL Erlenmeyer flask. Draw air in and out of the syringe enough times that you are certain that all the methanol has evaporated from it.

17. Collect temperature-pressure data for ethanol at room temperature. To do this:

a. Press [ENTER], then return to the main screen.

b. Repeat Steps 8-11 to do *one* trial only for ethanol in the room temperature water bath. When the pressure reading in Channel 2 stabilizes, record its value in the data table.

18. Open the 2-way valve of the pressure sensor. Remove the stopper assembly from the flask and dispose of the ethanol as directed by your instructor.

Processing the Data

1. Convert each of the Celsius temperatures to Kelvin (K). Record the answers.

2. To obtain the vapor pressure of methanol and ethanol, the air pressure must be subtracted from each of the measured pressure values. However, for Trials 2-4, even if *no* methanol was present, the pressure in the flask would have increased due to a higher temperature, or decreased due to a lower temperature (remember those gas laws?). Therefore, you must convert the atmospheric pressure at the temperature of the *first* water bath to a *corrected* air pressure at the temperature of the water bath in Trial 2, 3, or 4. To do this, use the gas-law equation (use the Kelvin temperatures):

$$\frac{P_2}{T_2} = \frac{P_1}{T_1} \qquad (15P.1)$$

where P_1 and T_1 are the atmospheric pressure and the temperature of the Trial 1 (room temperature) water bath. T_2 is the temperature of the water bath in Trial 2, 3, or 4. Solve for P_2, and record this value as the *corrected* air pressure for Trials 2, 3, and 4. For Trial 1 of methanol and Trial 1 of ethanol, it is not necessary to make a correction; for these two trials, simply record the atmospheric pressure value in the blank designated for air pressure.

3. Obtain the vapor pressure by subtracting the corrected air pressure from the measured pressure in Trials 2-4. Subtract the uncorrected air pressure in Trial 1 of methanol (and Trial 1 of ethanol) from the measured pressure.

4. Plot a graph of vapor pressure *vs.* temperature (°C) for the four data pairs you collected for methanol. Temperature is the independent variable and vapor pressure the dependent variable. As directed by your instructor, plot the graph manually or use Graphical Analysis software.

5. How would you describe the relationship between vapor pressure and temperature, as represented in the graph you made in the previous step? Explain this relationship using the concept of kinetic energy of molecules.

6. Which liquid, methanol or ethanol, had the larger vapor pressure value at room temperature? Explain your answer. Take into account various intermolecular forces in these two liquids.

Data and Calculations

Atmospheric pressure	____________ kPa

Substance	*Methanol*				*Ethanol*
Trial	1	2	3	4	1
Temperature (°C)	°C	°C	°C	°C	°C
Temperature (K)	K	K	K	K	K
Measured pressure	kPa	kPa	kPa	kPa	kPa

Air pressure	no correction kPa	corrected kPa	corrected kPa	corrected kPa	no correction kPa
Vapor pressure	kPa	kPa	kPa	kPa	kPa

Extension

The Clausius-Clapeyron equation describes the relationship between vapor pressure and absolute temperature:

$$\ln P = \Delta H_{vap} / RT + B$$

where ln P is the natural logarithm of the vapor pressure, ΔH_{vap} is the heat of vaporization, T is the absolute temperature, and B is a positive constant. If this equation is rearranged in slope-intercept form (y = mx + b):

$$\ln P = \Delta H_{vap} / R \cdot \frac{1}{T} + B$$

the slope, m, should be equal to $-\Delta H_{vap} / R$. If a plot of ln P *vs.* $1/T$ is made, the heat of vaporization can be determined from the slope of the curve. Plot the graph on the TI calculator:

1. Enter the temperature (°C) and vapor pressure values into your calculator:

TI-73 Calculators

a. To view the data lists, press [LIST].

b. Clear L1 and L2 by moving the cursor to the L1 or L2 headings and pressing [CLEAR] [ENTER].

c. Enter the four Celsius temperature values in L1 and the four corresponding *vapor pressure* values in L2. **Important:** Enter the values in order of increasing temperature.

d. To create a list of reciprocal of Kelvin temperature values (in L3), move the cursor until the L3 column heading is highlighted, then press [(] [2nd] [STAT] [L1] + 273 [)] [2nd] [X^{-1}] [ENTER].

e. To create a list of natural log (ln) of vapor pressure values (in L2), move the cursor until the L2 column heading is highlighted, press [MATH] [▶] [▶] [▶] and select [ln(] from the log menu. Then press [2nd] [STAT] [L2] [ENTER]. Proceed to Step 2.

TI-83 and TI-83 Plus Calculators

a. To view the data lists, press [STAT] to display the EDIT menu, and select Edit.

b. Clear L1 and L2 by moving the cursor to the L1 or L2 headings and pressing [CLEAR] [ENTER].

c. Enter the four Celsius temperature values in L1 and the four corresponding *vapor pressure* values in L2. **Important:** Enter the values in order of increasing temperature.

d. To create a list of reciprocal of Kelvin temperature values (in L3), move the cursor to the L3 column heading, then press [(] [2nd] [L1] + 273 [)] [x^{-1}] [ENTER].

e. To create a list of natural log (ln) of vapor pressure values (in L2), move the cursor to the L2 column heading, then press [LN] [2nd] [L2] [ENTER]. Proceed to Step 2.

TI-86 Calculators

a. To view the data lists, press `2nd` [STAT], then select <EDIT>.

b. Clear any previous data from lists L1 and L2—move the cursor to the L1 heading and press `CLEAR`, then `ENTER`. Do the same for the L2 list.

c. Enter the four temperature values in L1 and the four corresponding *vapor pressure* values in L2. **Important:** Enter the values in order of increasing temperature.

d. To create a list of reciprocal of Kelvin temperature values (in L3), move the cursor until the L3 column heading is highlighted, then select <NAMES>. Press `(` <L1> + 273 `)` `2nd` [x^{-1}] `ENTER`.

e. To create a list of natural log (ln) of vapor pressure values (in L2), move the cursor until the L2 column heading is highlighted, then press `LN` <L2> `ENTER`. Press `2nd` [QUIT]. Proceed to Step 2.

TI-89 Calculators

a. Press `APPS`, then select Home.

b. Enter the four Celsius temperature values in L1 and the four corresponding *vapor pressure* values in L2. **Important:** Enter the values in order of increasing temperature. First, enter the temperature values (shown here as t1, t2, ...) into L1. To do this, press `CLEAR` `2nd` [{] t1 `,` t2 `,` t3 `,` t4 `2nd` [}] `STO▶` `ALPHA` [L] `1` `ENTER`.

c. Now enter the four corresponding vapor pressure values (shown here as p1, p2, ...) into L2. To do this, press `CLEAR` `2nd` [{] p1 `,` p2 `,` p3 `,` p4 `2nd` [}] `STO▶` `ALPHA` [L] `2` `ENTER`.

d. To create a list of reciprocal of Kelvin temperature values (in L3), press `CLEAR` `1` `÷` `(` `ALPHA` [L] `1` + 273 `)` `STO▶` `ALPHA` [L] `3` `ENTER`.

e. To create a list of natural log (ln) of vapor pressure values (in L2), press `CLEAR` `LN` `ALPHA` [L] `2` `)` `STO▶` `ALPHA` [L] `2` `ENTER`. Proceed to Step 2.

TI-92 and TI-92 Plus Calculators

a. Press `APPS`, then select Home.

b. Enter the four Celsius temperature values in L1 and the four corresponding *vapor pressure* values in L2. **Important:** Enter the values in order of increasing temperature. First, enter the temperature values (shown here as t1, t2, ...) into L1. To do this, press `CLEAR` `2nd` [{] t1 `,` t2 `,` t3 `,` t4 `2nd` [}] `STO▶` `L` `1` `ENTER`.

c. Now enter the four corresponding vapor pressure values (shown here as p1, p2, ...) into L2. To do this, press `CLEAR` `2nd` [{] p1 `,` p2 `,` p3 `,` p4 `2nd` [}] `STO▶` `L` `2` `ENTER`.

d. To create a list of reciprocal of Kelvin temperature values (in L3), press `CLEAR` `1` `÷` `(` `L` `1` + 273 `)` `STO▶` `L` `3` `ENTER`.

e. To create a list of natural log (ln) of vapor pressure values (in L2), press [CLEAR] [LN] [L] [2] [)] [STO▶] [L] [2] [ENTER]. Proceed to Step 2.

2. Follow this procedure to calculate regression statistics and to plot a best-fit regression line on your graph of ln pressure *vs.* reciprocal of Kelvin temperature:

a. Start the DataMate program.

b. Select ANALYZE from the main screen.

c. Select CURVE FIT from the ANALYZE OPTIONS menu.

d. Select LINEAR (CH1 VS CH2) from the CURVE FIT menu. (CH1, or L2, is ln vapor pressure and CH2, or L3, is 1/Kelvin temperature.) The linear-regression statistics for these two lists are displayed for the equation in the form:

$$y = ax + b$$

where x is 1 / Kelvin temperature, y is ln vapor pressure, a is the slope, and b is the y-intercept. Record the value of the slope, a, to use in calculating the heat of vaporization in Step 3.

e. To display the linear-regression curve on the graph of ln vapor pressure *vs.* 1/Kelvin temperature (K), press [ENTER]. Examine your graph to see if the relationship between ln pressure vs. 1/ Kelvin temperature is linear.

f. (optional) Print a copy of the graph.

3. Use the slope value you recorded in Step 2 to calculate the heat of vaporization for methanol ($m = a = -\Delta H_{vap} / R$).

Freezing-Point Depression

EXP
16

Laboratory Time Required Three hours.

Special Equipment and Supplies

- Buret
- Buret clamp
- Freezing-point depression apparatus:
  - Large test tube
  - Two-hole stopper
  - Wire stirrer
  - Thermometer
- Tall-form beaker, 1000-mL
- Timer
- Solute[1]
- Ice
- Rock salt

Objective

In performing this experiment, students will determine the molar mass of an alcohol (Option A) or determine the extent of dissociation of citric acid in an aqueous solution of known concentration (Option B).

Safety

Although many alcohols are household chemicals, most are meant for external use only. Follow standard laboratory procedures: wear safety glasses and do not eat or drink in the lab.

Be careful not to break the thermometer or other glass apparatus. Clean up broken glass carefully to avoid being cut.

Consult your instructor about cleaning up spilled mercury.

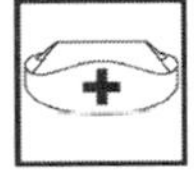

First Aid

If alcohol spills on your skin or enters your eyes, wash out the alcohol with copious amounts of water.

Seek medical attention if eye irritation results.

---

[1] Under Option A, the solute is an unknown alcohol. Under Option B, the solute is citric acid monohydrate.

Preamble

In this experiment, you will determine the molar mass of an alcohol or find the extent of dissociation of citric acid in aqueous solution.

Principles

Solutions of nonvolatile solutes have higher boiling points and lower freezing points than the solvents used to prepare the solutions. Both of these phenomena result from the fact that the vapor pressure of a solution (of a nonvolatile solute) is lower than the vapor pressure of the pure solvent.

It is relatively easy to see why the boiling point of a solution will be higher than that of the pure solvent. The vapor pressure of a liquid is determined by the ability of particles at the liquid's surface to escape into the vapor phase. In solutions of the type we are considering, some of the spaces at the surface are occupied by the nonvolatile solute particles. As a result, fewer solvent particles are in positions from which they can enter the vapor phase, and the solution's vapor pressure is lower than that of the pure solvent at all temperatures. This means that the solution will have to be heated to a higher temperature before its vapor pressure reaches that of the atmosphere. Hence, solutions of nonvolatile solutes have higher boiling points than do the pure solvents.

Although it may not seem as readily apparent, the same phenomenon (vapor pressure lowering) that causes the boiling-point elevation described above also leads to freezing-point depression. Figure 16.1 illustrates this effect.

The freezing point of a solution is the temperature at which the solvent in a solution and the pure solid solvent have the same vapor pressure. Vapor-pressure lowering results in freezing-point depression provided that the solution does not freeze as a solid solution. The solid that forms must be pure solvent.

Quantitatively, the magnitude of freezing-point depression is proportional to the amount of solute present in a given mass of solvent. The concentration unit employed is molality, symbolized by a lowercase m and defined as the number of moles of solute per kilogram of solvent. (See Equation 16.1.)

$$m = \frac{\text{moles solute}}{\text{kg solvent}} \tag{16.1}$$

The difference between the freezing point of the solution and that of the solvent is called the freezing-point depression and is given the symbol Δt_f. If Δt_f is calculated by subtracting the freezing temperature of the solvent from that of the solution, it will have a negative value. (See Equation 16.2.)

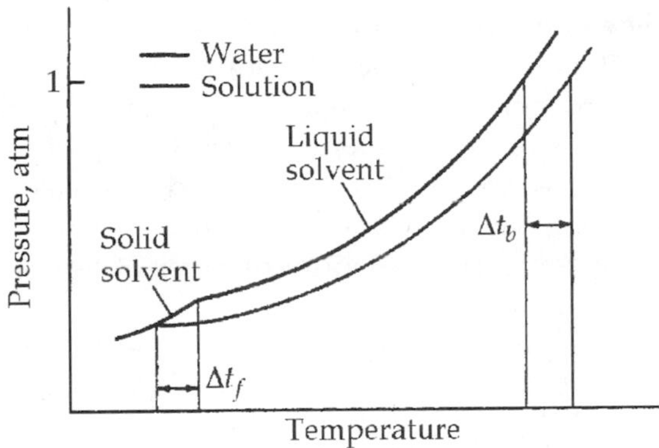

Figure16.1 Vapor pressure diagram for water and aqueous solutions

$$\Delta t_f = t_{f,\,\text{solution}} - t_{f,\,\text{solvent}} \tag{16.2}$$

The extent to which the freezing point is depressed for a solution of given molality is determined by the nature of the solvent. A 1.00 m aqueous solution would, ideally, have a freezing point of $-1.86°\text{C}$ because the freezing point of pure water is $0.00°\text{C}$. A 1.00 m solution in which camphor is the solvent would have a freezing point of $139°\text{C}$, which represents a depression by $40°\text{C}$ (the freezing point of pure camphor is $179°\text{C}$). Thus, the freezing-point depression constant, k_f, for water is $-1.86\text{K}/m$, while k_f for camphor has a value of $-40\text{K}/m$. The unit, K/m, is identical to the unit, $\text{K}\cdot\text{kg/mol}$ and to the units $°\text{C}/m$ and $°\text{C}\cdot\text{kg/mol}$ because temperature *differences* have the same magnitude on the Kelvin scale and on the Celsius scale. The relation between Δt_f, k_f, and m (for solutions of non-electrolytes) is shown in Equation 16.3.

$$\Delta t_f = k_f m \tag{16.3}$$

Equation 16.3 reveals that by measuring the freezing-point depression of a solution of a non-electrolyte, you can determine the solution's molality. Knowing this experimentally determined molality and knowing how the solution was prepared (i.e., what mass of solute was combined with what mass of solvent), you can obtain the molar mass of the solute. A computational scheme for obtaining the molar mass is shown in Equations 16.4 through 16.6.

$$m_{\text{exptl}} = \frac{\Delta t_f}{k_f} \tag{16.4}$$

$$\left(m_{\text{exptl}}\right)\left(\text{kg solvent}\right) = \text{moles solute} \tag{16.5}$$

$$\frac{\text{mass solute}}{\text{moles solute}} = \text{molar mass of solute} \tag{16.6}$$

In performing Option A of this experiment, you will determine the molar mass of an alcohol by measuring the freezing-point depression observed in an aqueous solution of the alcohol. This approach may be used very successfully, even though most alcohols are more volatile than water.

The experiment works because, at low temperatures, the vapor pressures of the alcohols are small enough to be considered negligible.

Equation 16.3 must be modified slightly for electrolytes, solutes that dissociate in aqueous solution. If the solute is a strong electrolyte, it may be considered to be completely dissociated in dilute solution and Equation 16.3 becomes Equation 16.7, in which the symbol ν (the Greek letter nu) represents the maximum number of ions obtained in the dissociation of a formula unit of the solute (e.g., $\nu = 2$ for NaCl while $\nu = 3$ for K_2SO_4).

$$\Delta t_f = \nu k_f m \tag{16.7}$$

Weak electrolytes may show considerably less than complete dissociation. For solutions of weak electrolytes, the relationship between the extent of freezing point depression and the molality is given by Equation 16.8 in which the symbol i has replaced the ν.

$$\Delta t_f = i k_f m \tag{16.8}$$

In solutions of solutes of known molar mass, one may determine the extent of dissociation of the solute by measuring the value of Δt_f and comparing it to extent of freezing-point depression that would be observed if the solute were, in fact, *not* dissociated. In performing Option B of this experiment, you will determine the value of i for a solution of citric acid.

Procedure

Procedure in a Nutshell

Prepare an ice/rock salt/water bath having a temperature no higher than −10°C. Use the bath to cool a sample of water and determine the freezing point of water. Prepare a solution of alcohol (Option A) or citric acid (Option B) in water. Cool the solution in the cold bath and determine its freezing point.

Preparing the Solutions

Prepare an ice/water/rock salt bath by placing equal amounts of ice and rock salt and a small amount of tap water in the 1000-mL tall-form beaker and stirring the mixture. Check the temperature of the bath and be sure it is at least as low as −10°C. If it is not, pour off some of the water, add more ice and rock salt, and stir the mixture again. Continue this process until the temperature of the bath is −10°C.

If you are performing Option A, make a solution of 10 mL of alcohol and 25 mL of water. Stir the mixture and transfer some to a test tube. Place the test tube in the rock salt/ice/water bath and observe whether the ice crystals begin to appear in the test tube; if they do, proceed to the determination of the freezing points of water and of the alcohol solution. If the alcohol mixture does not begin to freeze after several minutes in the cold bath, discard the mixture and prepare another solution, with 9 mL of alcohol in 25 mL of water. Proceed as before, decreasing the volume of alcohol in each subsequent trial mixture by 1 mL until you have a solution that will freeze in a cold bath maintained at −10°C.

If you are performing Option B, proceed as above, starting with 15 g of citric acid in 25 mL of water. If some of the solute does not dissolve, discard the mixture and prepare a new one, decreasing the mass of citric acid by one gram in each subsequent mixture until all of the solute dissolves. Once you have a solution (with no excess solute), see if it will freeze in the cold bath. If it will not, discard it and prepare a new solution, again decreasing the amount of solute by 1 gram until you have made a solution that will freeze in the cold bath.

Determination of the Freezing Points

Obtain a freezing-point depression apparatus like the one shown in Figure 16.2. If the thermometer has not been inserted into the stopper, insert it now. Most likely one side of the stopper will have been cut. The best way to insert the thermometer is to pull open the cut side of the stopper and place the thermometer inside the opening at an appropriate distance from the bottom of the thermometer.

Measure 25.0 mL of distilled water with the graduated cylinder and transfer it completely into the freezing-point depression test tube. Stopper the test tube with the stopper holding the wire stirrer and thermometer. Place the test tube in the ice/water/rock salt bath. Move the stirrer up and down constantly to agitate the water, pausing regularly, at 30-second intervals, to record the time and the temperature. Continue until the temperature readings have been constant for five consecutive times or until the water has frozen sufficiently that stirring has become difficult. Allow the frozen water to thaw and return to room temperature. Then repeat the procedure given above.

If you are performing Option A, rinse a clean buret with small amounts of alcohol and then fill the buret (to an appropriate height) with alcohol. Dispense a known volume of alcohol (recorded to the nearest 0.01 mL) into a pre-weighed flask. Weigh the flask and alcohol on the analytical balance. Use the data obtained to determine the alcohol's density.

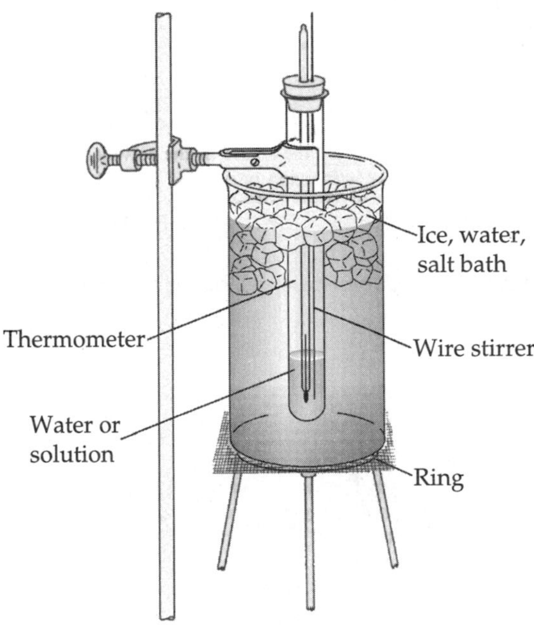

Figure 16.2 Freezing-point depression apparatus

After the water has thawed and returned to room temperature for a second time, (Option A) use the buret to dispense the appropriate volume of alcohol into the test tube or (Option B) weigh out the appropriate mass of citric acid monohydrate on the analytical balance and add it to the test tube of water. Record the exact volume of alcohol taken or mass of citric acid weighed out. Stopper the test tube containing the solution to be studied and place it in the bath. Repeat the process of agitation, recording the temperature every 30 seconds. Continue until five consecutive readings are the same or the mixture has frozen sufficiently that agitation becomes difficult. Allow the frozen mixture to thaw and return to room temperature. Then repeat the procedure.

If the water or solution does not freeze at a constant temperature, plot the time/temperature data. The points should fall on two straight lines (see Figure 16.3) with the intersection of the lines marking the freezing point of the material under study. The shaded portions of the plots in Figure 16.3 indicate supercooling, a phenomenon in which the liquid does not solidify until it is cooled to a temperature that is slightly below its freezing point.

Disposal of Reagents

Pour off the water from the bath and, if any solid rock salt remains, place the ice/rock salt mixture in styrofoam tubs for reuse by later classes. The mixture of isopropyl alcohol and water or of citric acid and water may be diluted and poured down the drain.

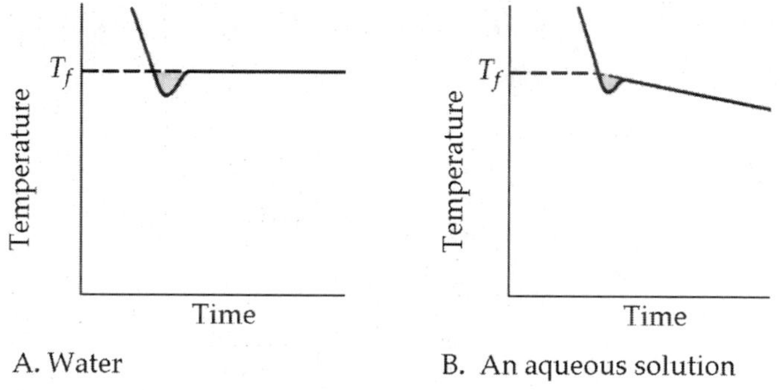

Figure 16.3 Cooling curves for (A) water and (B) an aqueous solution

name section date

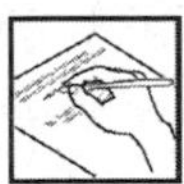

Pre-Lab Exercises for Experiment 16

These exercises are to be completed after you have read the experiment but before you come to the laboratory to perform it.

1. (*Option A*) A solution is prepared by mixing 2.17 g of an unknown nonelectrolyte with 225.0 g of chloroform. The freezing point of the mixture is –64.1°C, while the freezing point of pure chloroform is –63.5°C. The value of k_f for chloroform is –4.68 K/m. Find the molar mass of the unknown.

1. (*Option B*) A solution is prepared by mixing 0.6162 g of an magnesium sulfate heptahydrate, $MgSO_4 \cdot 7H_2O$ with 24.6 g of water. The freezing point of the mixture is –0.225°C, while the freezing point of pure water is 0.00°C. What is the value of i for this solution of magnesium sulfate?

2. Why is the freezing point of pure water determined experimentally rather than just assigned a value of 0°C?

name ______________________ section ______________________ date __________

Summary Report on Experiment 16

Time/Temperature Readings for Pure Water

Trial 1		Trial 2	
Time	*Temperature*	*Time*	*Temperature*

Time/Temperature Readings for Pure Water (cont.)

Trial 1		**Trial 2**	
Time	*Temperature*	*Time*	*Temperature*

name ______________________ section ______________________ date __________

Option A

Determination of Alcohol Density

Final buret reading ______________________

Initial buret reading ______________________

Volume of alcohol ______________________

Mass of flask plus alcohol ______________________

Mass of flask ______________________

Mass of alcohol ______________________

Density of alcohol ______________________

Preparation of Solution

Final buret reading ______________________

Initial buret reading ______________________

Volume of alcohol ______________________

Mass of alcohol ______________________

Option B

Mass of citric acid plus container ______________________

Mass of container ______________________

Mass of citric acid ______________________

Time/Temperature Readings for Mixture

Trial 1		Trial 2	
Time	*Temperature*	*Time*	*Temperature*

name　　section　　date

Time/Temperature Readings for Mixture (cont.)

Trial 1		Trial 2	
Time	*Temperature*	*Time*	*Temperature*

	Trial 1	Trial 2
Mass of solute used	________	________
t_f water	________	________
t_f mixture	________	________
Δt_f	________	________
m_{exptl}	________	________

Option A

	Trial 1	Trial 2
Moles alcohol in sample	________	________
Molar mass of alcohol	________	________
Average molar mass of alcohol	________	

Option B

	Trial 1	Trial 2
$k_f m$	________	________
i	________	________
i, average	________	

EXP
16P

Using Freezing-Point Depression to Find Molecular Weight

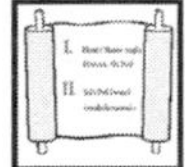

Principles

Solutions of nonvolatile solutes have higher boiling points and lower freezing points than the solvents used to prepare the solutions. Both of these phenomena result from the fact that the vapor pressure of a solution (of a nonvolatile solute) is lower than the vapor pressure of the pure solvent.

It is relatively easy to see why the boiling point of a solution will be higher than that of the pure solvent. The vapor pressure of a liquid is determined by the ability of particles at the liquid's surface to escape into the vapor phase. In solutions of the type we are considering, some of the spaces at the surface are occupied by the nonvolatile solute particles. As a result, fewer solvent particles are in positions from which they can enter the vapor phase, and the solution's vapor pressure is lower than that of the pure solvent at all temperatures. This means that the solution will have to be heated to a higher temperature before its vapor pressure reaches that of the atmosphere. Hence, solutions of nonvolatile solutes have higher boiling points than do the pure solvents.

Although it may not seem as readily apparent, the same phenomenon (vapor pressure lowering) that causes the boiling-point elevation described above also leads to freezing-point depression. Figure 16P.1 illustrates this effect.

The freezing point of a solution is the temperature at which the solvent in a solution and the pure solid solvent have the same vapor pressure. Vapor-pressure lowering results in freezing-point depression provided that the solution does not freeze as a solid solution. The solid that forms must be pure solvent.

Quantitatively, the magnitude of freezing-point depression is proportional to the amount of solute present in a given mass of solvent. The concentration unit employed is molality, symbolized by a lowercase m and defined as the number of moles of solute per kilogram of solvent. (See Equation 16P.1.)

$$m = \frac{\text{moles solute}}{\text{kg solvent}} \qquad (16P.1)$$

The difference between the freezing point of the solution and that of the solvent is called the freezing-point depression and is given the symbol Δt_f. If Δt_f is calculated by subtracting the freezing temperature of the solvent from that of the solution, it will have a negative value. (See Equation 16P.2.)

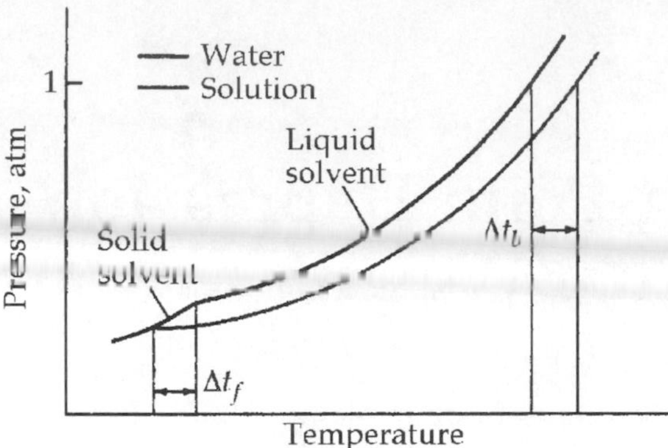

Figure 16P.1 Vapor pressure diagram for water and aqueous solutions

$$\Delta t_f = t_{f,\,\text{solution}} - t_{f,\,\text{solvent}} \tag{16P.2}$$

The extent to which the freezing point is depressed for a solution of given molality is determined by the nature of the solvent. A 1.00 m aqueous solution would, ideally, have a freezing point of –1.86°C because the freezing point of pure water is 0.00°C. A 1.00 m solution in which camphor is the solvent would have a freezing point of 139°C, which represents a depression by 40°C (the freezing point of pure camphor is 179°C). Thus, the freezing-point depression constant, k_f, for water is $-1.86\text{K}/m$, while k_f for camphor has a value of $-40\text{K}/m$. The unit, K/m, is identical to the unit, $\text{K}\cdot\text{kg/mol}$ and to the units $°\text{C}/m$ and $°\text{C}\cdot\text{kg/mol}$ because temperature *differences* have the same magnitude on the Kelvin scale and on the Celsius scale. The relation between Δt_f, k_f, and m (for solutions of non-electrolytes) is shown in Equation 16P.3.

$$\Delta t_f = k_f m \tag{16P.3}$$

Equation 16P.3 reveals that by measuring the freezing-point depression of a solution of a non-electrolyte, you can determine the solution's molality. Knowing this experimentally determined molality and knowing how the solution was prepared (i.e., what mass of solute was combined with what mass of solvent), you can obtain the molar mass of the solute. A computational scheme for obtaining the molar mass is shown in Equations 16P.4 through 16P.6.

$$m_{\text{exptl}} = \frac{\Delta t_f}{k_f} \tag{16P.4}$$

$$(m_{\text{exptl}})(\text{kg solvent}) = \text{moles solute} \tag{16P.5}$$

$$\frac{\text{mass solute}}{\text{moles solute}} = \text{molar mass of solute} \tag{16P.6}$$

In performing Option A of this experiment, you will determine the molar mass of an alcohol by measuring the freezing-point depression observed in an aqueous solution of the alcohol. This approach may be used very successfully, even though most alcohols are more volatile than water.

The experiment works because, at low temperatures, the vapor pressures of the alcohols are small enough to be considered negligible.

Equation 16P.3 must be modified slightly for electrolytes, solutes that dissociate in aqueous solution. If the solute is a strong electrolyte, it may be considered to be completely dissociated in dilute solution and Equation 16P.3 becomes Equation 16P.7, in which the symbol ν (the Greek letter nu) represents the maximum number of ions obtained in the dissociation of a formula unit of the solute (e.g., $\nu = 2$ for NaCl while $\nu = 3$ for K_2SO_4).

$$\Delta t_f = \nu k_f m \qquad (16P.7)$$

Weak electrolytes may show considerably less than complete dissociation. For solutions of weak electrolytes, the relationship between the extent of freezing point depression and the molality is given by Equation 16P.8 in which the symbol i has replaced the ν.

$$\Delta t_f = i k_f m \qquad (16P.8)$$

In solutions of solutes of known molar mass, one may determine the extent of dissociation of the solute by measuring the value of Δt_f and comparing it to extent of freezing point depression that would be observed if the solute were, in fact, *not* dissociated. In performing Option B of this experiment, you will determine the value of i for a solution of citric acid.

Materials

- CBL 2 interface
- TI Graphing Calculator
- DataMate program
- Temperature probe
- Ringstand
- Utility clamp
- 400-mL beaker
- 18 × 150-mL test tube
- Solute: Option A: unknown alcohol
  Option B: citric acid monohydrate
- Ice
- Rock salt
- Water

Procedure

1. Obtain and wear goggles.

2. Plug the temperature probe into Channel 1 of the CBL 2 interface. Use the link cable to connect the TI Graphing Calculator to the interface. Firmly press in the cable ends.

3. Turn on the calculator and start the DataMate program. Press CLEAR to reset the program.

4. Set up the calculator and interface for the temperature probe.

a. Select SETUP from the main screen.

b. If the calculator displays a temperature probe in CH 1, proceed directly to Step 5. If it does not, continue with this step to set up your sensor manually.

c. Press ENTER to select CH 1.

d. Select TEMPERATURE from the SELECT SENSOR menu.

e. Select the temperature probe you are using (in °C) from the TEMPERATURE menu.

5. Set up the data-collection mode.

   a. To select MODE, press ▲ once and press ENTER.

   b. Select TIME GRAPH from the SELECT MODE menu.

   c. Select CHANGE TIME SETTINGS from the TIME GRAPH SETTINGS menu.

   d. Enter "6" as the time between samples in seconds.

   e. Enter "100" as the number of samples. The length of the data collection will be 10 minutes.

   f. Select OK to return to the setup screen.

   g. Select OK again to return to the main screen.

Part I Preparing the Solutions

6. Prepare an ice/water/rock salt bath by placing equal amounts of ice and rock salt and a small amount of tap water in the 1000-mL tall-form beaker and stirring the mixture. Check the temperature of the bath and be sure it is at least as low as –10°C . If it is not, pour off some of the water, add more ice and rock salt, and stir the mixture again. Continue this process until the temperature of the bath is –10°C .

7. If you are performing Option A, make a solution of 10 mL of alcohol and 25 mL of water. Stir the mixture and transfer some to a test tube. Place the test tube in the rock salt/ice/water bath and observe whether the ice crystals begin to appear in the test tube; if they do, proceed to the determination of the freezing points of water and of the alcohol solution.

8. If the alcohol mixture does not begin to freeze after several minutes in the cold bath, discard the mixture and prepare another solution, with 9 mL of alcohol in 25 mL of water. Proceed as before, decreasing the volume of alcohol in each subsequent trial mixture by 1 mL until you have a solution that will freeze in a cold bath maintained at –10°C .

9. If you are performing Option B, proceed as above, starting with 15 g of citric acid in 25 mL of water. If some of the solute does not dissolve, discard the mixture and prepare a new one, decreasing the mass of citric acid by one gram in each subsequent mixture until all of the solute dissolves. Once you have a solution (with no excess solute), see if it will freeze in the cold bath. If it will not, discard it and prepare a new solution, again decreasing the amount of solute by 1 gram until you have made a solution that will freeze in the cold bath.

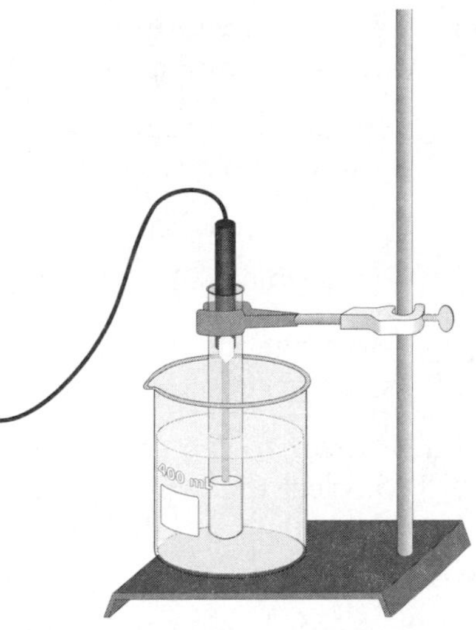

Figure 16P.2 Apparatus setup

Part II Determination of the Freezing Points

10. Obtain a freezing-point depression apparatus like the one shown in Figure 16P.2. If the temperature probe has not been inserted into the stopper, insert it now. Most likely one side of the stopper will have been cut. The best way to insert the temperature probe is to pull open the cut side of the stopper and place the temperature probe inside the opening at an appropriate distance from the bottom of the test tube.

11. Measure 25.0 mL of distilled water with the graduated cylinder and transfer it completely into the freezing-point depression test tube. Stopper the test tube with the stopper holding the wire stirrer and temperature probe.

12. After 30 seconds have elapsed, select START to begin data collection.

13. Place the test tube in the ice/water/rock salt bath. Make sure the water level outside the test tube is higher than the solution level inside the test tube.

14. With a slight up and down motion of the stirrer, *continuously* stir the solution during the cooling.

15. Continue until the water has frozen sufficiently that stirring has become difficult. Data collection will stop after 10 minutes. Allow the frozen water to thaw and return to room temperature. Then repeat the procedure given above.

16. To determine the freezing temperature of water, you need to determine the temperature in the portion of the graph with nearly constant temperature. Examine the data points along this portion of the graph. As you move the cursor right or left, the time (X) and temperature (Y) values of each data point are displayed below the graph. Determine this temperature, either by visually approximating the value or by taking the mathematical average of the temperatures in this plateau. Record the freezing temperature of water in your data table (round to the nearest 0.1°C).

17. To determine the freezing temperature of water, you need to analyze the portion of the graph with nearly constant temperature. To do this:

a. Press ENTER to return to the main screen, then select ANALYZE.

b. Select STATISTICS from the ANALYZE OPTIONS menu.

c. Use ▶ to move the cursor to the beginning of the flat section of the curve. Press ENTER to select the left boundary of the flat section.

d. Move the cursor to the end of the flat section of the graph, and press ENTER to select the right boundary of the flat section. The program will now calculate and display the statistics for the data between the two boundaries.

e. Record the MEAN value as the freezing temperature in your data table (round to the nearest 0.1°C).

f. Press ENTER to return to the ANALYZE OPTIONS menu, then select RETURN TO MAIN SCREEN.

18. Store the data from the first run so that it can be used later. To do this:

a. Select TOOLS from the main screen.

b. Select STORE LATEST RUN from the TOOLS MENU.

Part III Freezing Temperature of an Alcohol or a Solution of Citric Acid

19. If you are performing Option A, rinse a clean buret with small amounts of alcohol and then fill a buret (to an appropriate height) with alcohol. Dispense a known volume of alcohol (rounded to the nearest 0.01 mL) into a pre-weighed flask. Weigh the flask and alcohol on the analytical balance. Use the data obtained to determine the alcohol's density.

20. After the water has thawed and returned to room temperature for second time, (Option A) use the buret to dispense the appropriate volume of alcohol into the test tube or (Option B) weigh out the appropriate mass of citric acid monohydrate on the analytical balance and add it to the test tube of water.

21. Record the exact volume of alcohol taken or mass of citric acid weighed out. Stopper the test tube containing the solution to be studied and place it in the bath.

22. Choose START to begin data collection.

23. Repeat the process of agitation, recording the temperature every 30 seconds. Continue until five consecutive readings are the same or the mixture has frozen sufficiently that agitation becomes difficult. Allow the frozen mixture to thaw and return to room temperature. Then repeat the procedure.

24. If the water or solution does not freeze at a constant temperature, plot the time/temperature data. The points should fall on two straight lines with the intersection of the lines marking the freezing point of the material under study.

25. To determine the freezing point of the solution, you need to determine the temperature at which the mixture first started to freeze. Examine the data points to locate the freezing point of the solution, as shown in Figure 16P.3. As you move the cursor right or left on the displayed graph, the time (X) and temperature (Y) values of each data point are displayed below the graph. Record the freezing point in the Data and Calculations table (round to 0.1°C).

26. A good way to compare the freezing curves of the pure substance and the mixture is to view both sets of data on one graph. To do this:

a. Press ENTER to return to the main screen.

b. Select GRAPH from the main screen, then press ENTER.

c. Select MORE, then select L2 AND L3 VS L1 from the MORE OPTIONS menu.

d. Both temperature runs should now be displayed on the same graph.

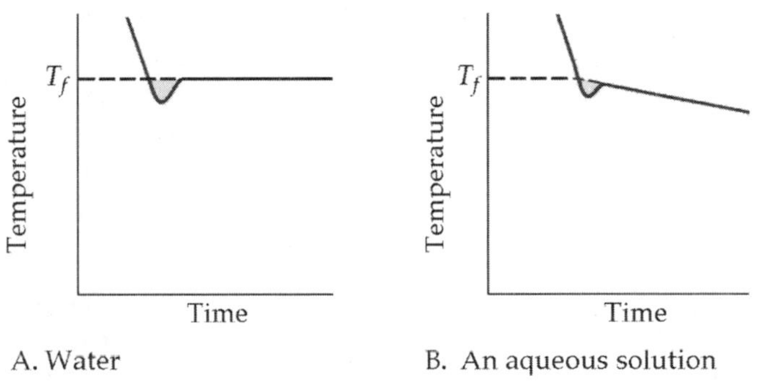

Figure 16P.3 Cooling curves for (A) water and (B) an aqueous solution

name section date

Data and Calculations

Option A:

Determination of Alcohol Density

Final buret reading	mL
Initial buret reading	mL
Volume of alcohol	mL
Mass of flask plus alcohol	g
Mass of flask	g
Mass of alcohol	g
Density of alcohol	g/mL

Preparation of Solution

Final buret reading	mL
Initial buret reading	mL
Volume of alcohol	mL
Mass of alcohol	g

Option B:

Mass of citric acid plus container	g
Mass of container	g
Mass of citric acid	g

	Trial 1	*Trial 2*
Mass of solute used	g	g
t_g water		
t_f mixture		
Δt_f		
m_{exptl}		

Option A:

Moles alcohol in sample		
Molar mass of alcohol		
Average molar mass of alcohol		

Option b:

$k_f m$		
i		
I, average		

EXP
17

Absorption Spectroscopy and Beer's Law

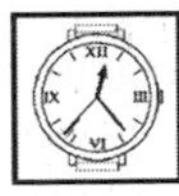

Laboratory Time Required Three hours.

Special Equipment and Supplies

Spectrophotometer
Buret or Mohr pipet
Volumetric flask, 100-mL
Pipet bulb
Spectrophotometer cells

Cobalt(II) chloride hexahydrate, $CoCl_2 \cdot 6H_2O$
Copper(II) sulfate pentahydrate, $CuSO_4 \cdot 5H_2O$
Nickel(II) nitrate hexahydrate, $Ni(NO_3)_2 \cdot 6H_2O$

Objective In performing this experiment, the student will be introduced to the use of spectrophotometry in chemical analysis.

Safety

The $CoCl_2 \cdot 6H_2O$ and $Ni(NO_3)_2 \cdot 6H_2O$ are classified as poisons and, consequently, should not be ingested or discarded in the sanitary sewer.

Clean up spills promptly and avoid exposing your skin to the solid reagents or their solutions.

Safety goggles should be worn at all times. Laboratory aprons are strongly advised.

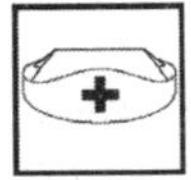

First Aid

In case of contact with the solid reagents or their solutions, rinse the exposed area thoroughly with water.

Preamble

Light brings us news of the universe. Our perception of light is mostly of the visible range of radiant energies, the slot between roughly 400 and 800 nm within the electromagnetic spectrum. But because most light is invisible, it must be detected by other means. Through the transmission, absorption, and scattering of light, we can gain information about the hidden world beyond what usual observations reveal.

In this experiment, you will study the Beer-Lambert Law, one of the most fundamental and widely applied spectroscopic laws. In the first part of the experiment, you will determine where in the visible range of the electromagnetic spectrum a solution of a transition metal salt absorbs light. In the second part, you will determine the nature of the relationship between absorbance and the concentration of the solution. In the third part of the experiment, you will study a mixture of the solutions and determine how the appearance of the mixture relates to its spectrum.

Principles

Spectroscopy is the study of the interaction of electromagnetic radiation with matter. In spectroscopy, two terms are inescapable: transmittance and absorbance. **Transmittance** (T) is defined as the ratio of the intensity of light after it passes through the medium being studied (I) to the intensity of light before it encountered the medium (I_0), as shown in Equation 17.1.

$$T = I / I_0 \qquad (17.1)$$

Spectroscopists more commonly refer to percent transmittance (%T), which is simply: $I/I_0 \times 100\%$. Often the same spectroscopic information that is reported as the percent transmittance is more conveniently expressed as absorbance (A):

$$A = -\log(I / I_0) \qquad (17.2)$$

Note that

$$A = 2 - \log(\%T) \qquad (17.3)$$

If one knows the percent transmittance, one can calculate absorbance and vice versa. Some non-digital spectrophotometers have both a %T and an absorbance scale displayed on a meter. Because the %T scale is linear, it can be read with good precision over the entire range of transmittances. However, the absorbance scale is a logarithmic scale and cannot be read with precision at high absorbance values. Therefore, if the absorbance is larger than 0.7, it is preferable to calculate the absorbance, using the %T, rather than to read the absorbance directly. Modern digital spectrophotometers may be programmed to display either absorbance or percent transmittance by simply touching a mode selection button.

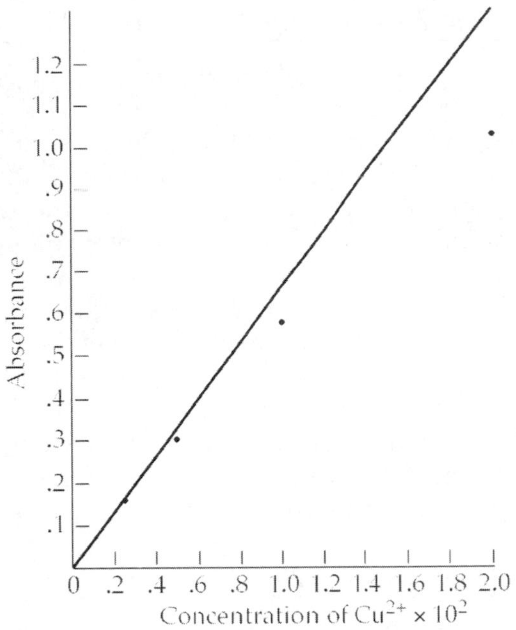

Figure 17.1 Cu^{2+}/NH_3 solutions do not obey Beer's Law. A proper calibration curve would be a smooth line drawn through the data points.

An operational statement of the Beer-Lambert Law can be represented as

$$A = \varepsilon c \ell \tag{17.4}$$

where c is the concentration of some absorbing substance in solution, ℓ is the optical path length, and ε is the molar absorptivity. The molar absorptivity is a constant that depends on the nature of the absorbing system (the solute–solvent combination) and the wavelength of the light passing through it. A plot that shows the dependence of A (or ε) on wavelength is called a **spectrum**. In one part of this experiment, you will determine the visible spectrum of an aqueous solution of cobalt chloride, copper sulfate, or nickel nitrate.

When absorbance measurements are made at a fixed wavelength in a cell of constant path length, ε and ℓ are constant. Therefore, the absorbance, A, should be directly proportional to c, the concentration of the solute. A solution that shows such a linear relation between A and c is said to obey the Beer-Lambert Law, which is the optimum situation for a spectrophotometric method of analysis. In the second part of this experiment, you will attempt to verify whether some or all the aqueous solutions studied obey the Beer-Lambert Law.

It is also possible, although less desirable, to perform spectrophotometric analyses in systems that deviate from the Beer-Lambert Law, as the Cu^{2+}/NH_3 system is shown to do in Figure 17.1.

The deviation results from the equilibrium distribution of Cu^{2+} between the absorbing species $Cu(NH_3)_4^{2+}$ and several nonabsorbing complexes of general formula $Cu(NH_3)_{4-i} \cdot (H_2O)_i^{2+}$. In such cases, the absorbance versus concentration measurements result in a calibration curve, rather than the linear plot predicted by the Beer-Lambert Law.

Procedure

Procedure in a Nutshell

Find the colors of visible light associated with its various wavelengths. Prepare a solution of known concentration and determine its spectrum. Prepare various dilute solutions and measure their absorbances at a fixed wavelength in order to prepare a Beer's Law plot. Obtain the spectrum of a mixture of various colored salts.

A. Operation of the Spectrophotometer

The large variety of spectrometers available in undergraduate teaching laboratories makes it impossible for us to give operating instructions for each type. Your instructor will demonstrate appropriate techniques for the instrument you will use in this experiment.

You will be asked to determine the wavelengths of monochromatic light that correspond to certain colors of the visible spectrum. In some instruments you may accomplish this task by placing a spectrophotometer cell containing a 3/4-inch piece of chalk in the sample compartment, and slowly changing the wavelength over the 400- to 700-nm range while looking into the test tube. You may need to raise, lower, or turn the test tube to obtain the best reflection from the chalk surface. In other spectrophotometers, you may be able to hold a lint-free tissue in the path of the light while changing the wavelength; the color of the light should be clearly visible on the tissue. Record your observations on the Summary Report Sheet.

B. Preparation of 0.200 M Aqueous Salt Solution

Obtain a salt assignment from your instructor. Calculate the mass of your assigned salt needed to prepare 100.00 mL of a 0.200 M solution of your salt and weigh the necessary mass on the analytical balance. Place your sample in a clean beaker and add about 50 mL of water. Swirl the beaker until the solid has totally dissolved; then carefully pour the solution into a 100-mL volumetric flask. Rinse the beaker with several small volumes of water and add each rinsing to the solution in the flask. Finally, dilute the solution in the volumetric flask to exactly 100 mL, and mix well.

C. Determination of the Spectrum of the Salt Solution

Obtain two matched spectrophotometer cells and fill one of them half full with distilled water to use as a reference. Set the wavelength of the spectrophotometer at 390 nm; then insert the cell in the cell compartment, taking care to align the reference marks. Close the compartment door and set the instrument to exactly 100%*T* or zero absorbance. Remove the reference cell and adjust the zero setting, if necessary. (Note: Some modern spectrophotometers do not require manual adjustment of the zero transmittance point.) Check the zero and 100%*T* (or zero absorbance) settings until they are reproducible. Next rinse the sample cell with a small portion of the 0.200 M solution and discard the rinsing. Fill the cell about two-thirds full with solution, insert it in the sample compartment, and align the reference marks. Close the sample door and read the %*T* (or absorbance) as accurately as possible.

Change the wavelength to 400 nm and again adjust the zero and 100% *T* (or zero absorbance) settings using the reference cell as described in the previous paragraph. Replace the reference cell with the sample cell and record the %*T* (or absorbance) of the solution. Repeat this procedure at 10-nm intervals over the 390 to 600 nm range.

Prepare a plot of absorbance versus wavelength, using Equation 17.3 if necessary, to convert from %*T* to absorbance. Determine the wavelength of maximum absorbance for your salt. Obtain the absorbance maxima for the other salts from your classmates who studied them.

D. Beer's Law

Place four clean, dry test tubes in a test tube block. Use a buret or Mohr pipet to dispense 6 mL, 4 mL, 2 mL, and 1 mL volumes of the your 0.200 M solution into the four test tubes. In the same way, add 2, 4, 6, and 7 mL volumes of distilled water, respectively, to the test tubes, and swirl or thump the tubes to mix the solutions. Each test tube should now contain 8 mL of solution. Calculate the concentration of your assigned salt in each test tube.

Set the spectrophotometer wavelength to the absorbance maximum and adjust the zero and 100%*T* (or zero absorbance) setting, using the reference cell as usual. Rinse and then fill the sample cell with one of the diluted cobalt chloride solutions and measure the %*T* (or absorbance) of the solution. Record your data on the Summary Report Sheet. Repeat this procedure for each of the your diluted solutions. Plot absorbance (obtained from %*T* via Equation 17.3 if necessary) versus the concentration of your salt for the five solutions (including the original 0.200 M solution). If the data display a linear relationship, draw the best straight line through the experimental points, including the origin.

Obtain absorbance (or %*T*) and concentration data from classmates who studied the other salts and prepare plots of these data as well. Fit the best straight line to each set of data to determine its linearity and comment on whether the relationship between absorbance and concentration is linear or not.

E. Determination of the Spectrum of a Mixture

Work with classmates who studied salts other than the one you studied to prepare a mixture specified by your instructor. Note the color of your mixture on the Summary Report Sheet and obtain its spectrum by the procedure specified in Procedure C.

Disposal of Reagents

Any solutions containing cobalt or nickel ions should be placed in labeled collection bottles.

Questions

1. If the solution obeys the Beer-Lambert Law, why is it better to plot the absorbance versus concentration data for several concentrations, rather than using only a single solution plus the origin to determine the straight line?

2. What does it mean if the experimental points obviously curve away from the expected straight line?

3. Calculate the molar absorptivity, at the absorbance maximum, for each of the salts using a path length of 1/2 inch (1.27 cm) for test tube spectrophotometer cells or 1.00 cm for square spectrophotometer cells.

4. Which of the salts, if any, give spectra that do not obey Beer's Law?

5. Although Cu^{2+} / NH_3 solutions do not obey Beer's Law, they are usually used in preference to simple aqueous solutions of Cu^{2+} when copper ion concentration is being determined spectrophotometrically. Explain why.

name section date

Pre-Lab Exercises for Experiment 17

These exercises are to be completed after you have read the experiment but before you come to the laboratory to perform it.

1. Why is it necessary to recalibrate the spectrophotometer against a reference cell each time the wavelength is changed?

2. If a solution measures 75.3%*T*, what is the absorbance of the solution?

3. A solution of K_2CrO_4 has an absorbance of 0.395 in a cell of 1.00 cm pathlength at a wavelength of 370 nm. The value of ε (the molar absorptivity) is 4.84×10^3 M^{-1} cm^{-1} for K_2CrO_4 at that wavelength. What is the concentration of the solution?

name ______ section ______ date ______

Summary Report on Experiment 17

A. Operation of the Spectrophotometer

Light Transmitted	**Wavelength, nm**
Red	______
Orange	______
Yellow	______
Green	______
Blue	______
Violet	______

B. Preparation of 0.200 M Aqueous Solution

Assigned salt	______
Mass of beaker plus salt	______
Mass of beaker	______
Mass of salt	______

C. Determination of the Spectrum of Your 0.200 M Solution

Wavelength	390	400	410	420	430	440	450
% Transmittance	____	____	____	____	____	____	____
Absorbance	____	____	____	____	____	____	____

Wavelength	460	470	480	490	500	510	520
% Transmittance	____	____	____	____	____	____	____
Absorbance	____	____	____	____	____	____	____

Wavelength	530	540	550	560	570	580	590	600
% Transmittance	____	____	____	____	____	____	____	____
Absorbance	____	____	____	____	____	____	____	____

What is the wavelength of the absorbance maximum? ______

Other salts

Name of salt	______	______
Obtained from	______	______
Mass used in preparing solution	______	______
Wavelength of maximum absorbance	______	______

D. Beer's Law

Salt	______				
Wavelength used	______				
Concentration (M)	0.200	____	____	____	____
% Transmittance	____	____	____	____	____
Absorbance	____	____	____	____	____

Salt	______				
Wavelength used	______				
Concentration (M)	0.200	____	____	____	____
% Transmittance	____	____	____	____	____
Absorbance	____	____	____	____	____

Salt	______				
Wavelength used	______				
Concentration (M)	0.200	____	____	____	____
% Transmittance	____	____	____	____	____
Absorbance	____	____	____	____	____

name ______________________ section ______________________ date __________

E. *Determination of the Spectrum of a Mixture*

Specifications for the mixture ______________________

____

Appearance of the mixture ______________________

____

Wavelength	390	400	410	420	430	440	450
% Transmittance	____	____	____	____	____	____	____
Absorbance	____	____	____	____	____	____	____

Wavelength	460	470	480	490	500	510	520
% Transmittance	____	____	____	____	____	____	____
Absorbance	____	____	____	____	____	____	____

Wavelength	530	540	550	560	570	580	590	600
% Transmittance	____	____	____	____	____	____	____	____
Absorbance	____	____	____	____	____	____	____	____

Is the spectrum of the mixture what you would have expected based on its color? Explain your answer briefly.

____

____

EXP
18

Getting the End Point to Approximate the Equivalence Point

Laboratory Time Required

Three hours.

Special Equipment and Supplies

Balance
Burets
Buret clamp
pH electrode
pH meter
Potassium hydrogen phthalate, KHP(*s*)
0.1 M sodium hydroxide, NaOH (*aq*)
0.1 M hydrochloric acid, HCl (*aq*)
Phenolphthalein
Universal Indicator
pH = 7 buffer
pH = 4 buffer

Objective

In performing this experiment, the student will consider the factors that affect an indicator's suitability for use in an acid/base titration.

Safety

Bases, such as sodium hydroxide, can cause skin burns and are especially hazardous to the eyes. Acids, such as hydrochloric acid, can also cause skin burns.

Safety glasses with side shields are required.

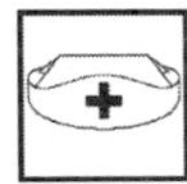

First Aid

Following skin contact with either sodium hydroxide or hydrochloric acid, wash the area thoroughly with water.

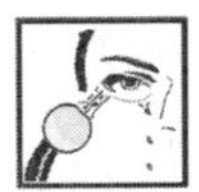

Should either solution get in the eyes, rinse them with thoroughly with water. At least 20 minutes of flushing with water is recommended. Then seek medical attention.

Preamble

There are a number of experiments in this manual that involve acid/base indicators. In Experiment 8, phenolphthalein is used to mark the end point in the titration of vinegar with sodium hydroxide. Methyl orange serves the same purpose in the titration of hydrochloric acid with ammonia (Experiment 23). The acidity constants of indicators are studied in Experiment 19. In this experiment we explore the question of why different indicators change color at different pH values and illustrate that the end point of a titration (signaled by an indicator's color change) does not necessarily correspond to the equivalence point in the titration.

Principles

The equivalence point in a titration is the point at which the acid and base have been mixed in "stoichiometric proportion," meaning that neither acid nor base is in excess. For example, if 50.00 mL of 0.1000 M hydrochloric acid is titrated with 0.1000 M sodium hydroxide, the equivalence point will be reached when 50.00 mL of base have been added to the acid. At that point in the titration, the acid and base have exactly neutralized each other. The question is "How can one determine that this point has been reached?"

In the early portion of the titration, the chemical amount of base that has been added to the titration mixture is less than the chemical amount of acid present and the pH of the titration is less than 7 (at 25°C). At the equivalence point, the titration mixture is essentially a 0.05000 M solution of sodium chloride and should have a pH of 7.00 (at 25°C). Further addition of base raises the pH above 7 (at 25°C) because there is no longer any acid present to react with the extra sodium hydroxide. These changes in pH take place without any apparent changes in the appearance of the reaction mixture. The equivalence point is revealed either by monitoring the changing pH of the solution (by the use of a pH meter) or by finding an indicator that will change color in the vicinity of the equivalence point. However, not all indicators will change color at the desired point in the titration.

Indicators

As discussed in Experiment 17, indicators are weak acids that have different colors at low pH values (when the indicator is predominantly in its HIn form) and at high pH values (when the indicator is predominantly in its In^- form). A good rule of thumb is that the indicator will display its low pH color when the ratio of [HIn] to $[In^-]$ has a value of 10 or more; conversely, the indicator will display its high pH color when the ratio of [HIn] to $[In^-]$ has a value of 0.1 or less. Consider an indicator that is blue in the HIn form and is red in the In^- form. If the indicator has a K_a value of 1.0×10^{-4}, it will appear blue at pH's below 3 and red at pH's above 5. (See Equations 18.1 through 18.3.)

$$K_a = \frac{[H_3O^+][In^-]}{[HIn]} \tag{18.1}$$

$$1.0\times10^{-4} = \frac{[H_3O^+](1)}{(10)};\ [H_3O^+] = 1.0\times10^{-3} \text{ when } [HIn] = 10[In^-] \tag{18.2}$$

$$1.0\times10^{-4} = \frac{[H_3O^+](10)}{(1)}); [H_3O^+] = 1.0\times10^{-5} \text{ when } [In^-] = 10[HIn] \qquad (18.3)$$

Calculations similar to those shown in Equations 18.1 through 18.3 reveal that, if the indicator had a K_a value of 1.0×10^{-8}, it would appear blue at pH's below 7 and red at pH's above 9. Thus, an indicator with $K_a = 1.0\times10^{-4}$ would change color before the equivalence point was reached in the titration of HCl by NaOH while an indicator with $K_a = 1.0\times10^{-8}$ would change color after the equivalence point had been reached in the same titration.

Universal Indicator

Universal Indicator is actually a mixture of various indicators, chosen so that the mixture will undergo several color changes as the pH of the solution being titrated varies from a value of 4 to a value of 10. In this experiment, you will perform a pH titration of hydrochloric acid by sodium hydroxide in the presence of Universal Indicator. This will permit you to determine whether a given color change occurs near the equivalence point or not.

Procedure

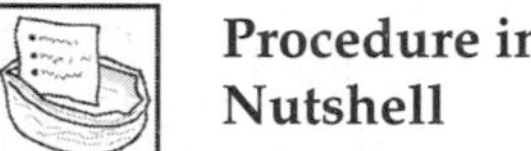

Procedure in a Nutshell

Standardize 0.1 M NaOH via titration with KHP, using phenolphthalein as an indicator. Standardize 0.1 M HCl via titration with the NaOH, again using phenolphthalein as an indicator. Add 3 drops of Universal Indicator to 25 mL of hydrochloric acid and titrate it with NaOH, monitoring the course of the titration with a pH meter.

Standardizations

Clean a buret and prepare it for use in the standardization of sodium hydroxide according to the directions provided in the Introduction. Accurately weigh, to the nearest 0.1 mg, 0.4 g of KHP. Place the KHP in an Erlenmeyer flask, dissolve it in 25 mL of distilled water, and add 2 to 3 drops of phenolphthalein. Read and record the initial volume of NaOH in the buret to the nearest 0.01 mL. Titrate the KHP solution until a faint pink color, which does not disappear when the solution is swirled, is obtained. Read and record the final volume of NaOH in the buret to the nearest 0.01 mL.

Refill the buret with sodium hydroxide. Clean another buret and prepare it for use in the standardization of hydrochloric acid according to the directions provided in the Introduction. Deliver a 25 mL HCl sample to an Erlenmeyer flask (read and record the initial and final buret readings to the nearest 0.01 mL). Add 2 to 3 drops of phenolphthalein to the acid. Read and record the initial volume of NaOH in the buret to the nearest 0.01 mL. Titrate the HCl solution to the phenolphthalein end point. Read and record the final volume of NaOH in the buret to the nearest 0.01 mL.

Calculate the molarities of the NaOH and the HCl solutions. Determine the volume of NaOH that will have to be added to 25 mL of HCl to reach the equivalence point.

pH Titration

The large variety of pH meters available in undergraduate teaching laboratories makes it impossible for us to give operating instructions for each. Your instructor will provide directions for calibrating your pH meter.

Refill the base buret with sodium hydroxide and the acid buret with hydrochloric acid. Deliver a 25 mL HCl sample to an Erlenmeyer flask (read and record the initial and final buret readings to the nearest 0.01 mL). Add 2 to 3 drops of Universal Indicator to the acid. Read and record the initial volume of NaOH in the buret to the nearest 0.01 mL. Rinse the electrode with distilled water and shake it to dry the bulb. Immerse the electrode in the HCl solution. Move the pH meter's Function Knob to "pH." Record the initial pH of the HCl.

Begin the titration. Start by adding 1–2 mL increments of NaOH. Swirl the acid flask and record the pH meter reading after each addition of NaOH. Note the pH of the titration mixture at which each color change occurs.

As the volume of base added approaches the calculated equivalence point, decrease the size of the NaOH increment. (Add the NaOH dropwise for 2 mL before and 2 mL after the anticipated equivalence point.) Continue adding NaOH, swirling the flask, and recording the pH (and any color change) until you have gone 10 mL beyond the equivalence point.

Prepare a plot of pH (*y*-axis) versus volume of NaOH added (*x*-axis). The equivalence point is marked by a large increase in the pH upon the addition of a small volume of base.

Calculations

Titrations are usually performed to find the molarity of an acid or base. Assume that the molarity of your HCl solution is correct and calculate the molarity of NaOH (using $M_aV_a = M_bV_b$) using the volume of base added that corresponds to (1) each of the observed color changes and (2) the equivalence point as determined on the pH plot. Calculate the % difference between the molarity based on the pH titration and the molarity based on each color change. State which indicators should be used to guarantee that the end point of the titration would be an accurate approximation of the equivalence point in the titration of HCl by NaOH.

Disposal of Reagents

Excess KHP should be placed in the containers used for solid waste. Solutions should be neutralized and diluted. They may then be flushed down the drain.

name section date

Pre-Lab Exercises for Experiment 18

These exercises are to be completed after you have read the experiment but before your come to the laboratory to perform it.

1. Consider the titration of 25.00 mL of 0.2000 M hydrochloric acid by 0.2000 M sodium hydroxide. Calculate the pH of the mixture at the various points in the titration listed below.
   a. start (no NaOH added)

   b. after the addition of 10.00 mL of NaOH

   c. after the addition of 24.50 mL of NaOH

   d. at the equivalence point (after the addition of 25.00 mL of NaOH)

   e. after the addition of 25.50 mL of NaOH

2. Would an indicator with $K_a = 10^{-3}$ be a suitable indicator for the titration described in Question 1?

name ______ section ______ date ______

Summary Report on Experiment 18

Standardization of NaOH Solution

Mass of KHP and container ______

Mass of container ______

Mass of KHP ______

Final buret reading, NaOH ______

Initial buret reading, NaOH ______

Volume used, NaOH ______

Molarity of NaOH solution ______

Standardization of HCl Solution

Final buret reading, HCl ______

Initial buret reading, HCl ______

Volume used, HCl ______

Final buret reading, NaOH ______

Initial buret reading, NaOH ______

Volume used, NaOH ______

Molarity of HCl solution ______

pH Titration

Buret reading, mL	*Volume of base added, mL*	*pH*	*Color*

name ______ section ______ date ______

Buret reading, mL	Volume of base added, mL	pH	Color

EXP
19

Determination of the Dissociation Constant of an Acid/Base Indicator

Laboratory Time Required Three hours.

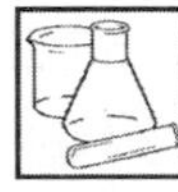

Special Equipment and Supplies

- Spectrophotometer
- Sample cells (cuvets)
- Buret
- Mohr pipets
- Buffer solutions, varying in pH from 3.5 to 5.5
- 1 M $HCl(aq)$, hydrochloric acid
- 1 M $NaOH(aq)$, sodium hydroxide
- Indicator solutions

Objective In this experiment, students will use spectrophotometric data to find the value of the acidity constant, K_a, of an indicator (a weak acid that changes color upon deprotonation).

Safety

Acids and bases are harmful to the skin and eyes. Avoid contact with acids and bases.

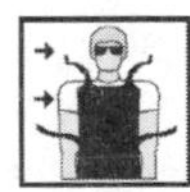

Safety goggles are required; laboratory aprons are advised.

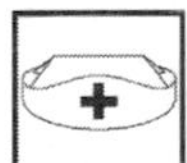

First Aid

Wash off any acids or bases spilled on the skin with copious amounts of water.

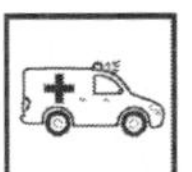

Should acid or base get in your eyes, flush the chemicals out by washing your eyes with water for at least 20 minutes. Seek medical attention.

Preamble

An acid/base indicator is a weak acid or weak base whose protonated (HIn) and deprotonated (In^-) forms have different colors in aqueous solution. This difference in colors is obviously necessary for the substance to be used as a visual indicator. Because the different colors result from the two forms of the indicator differing in their ability to absorb specific wavelengths of light, the absorption of light by an indicator solution may be used to study its ionization reaction. In this experiment, you will use absorption spectrophotometry to measure the dissociation constant for one of several indicators.

Principles

The acid/base indicators to be studied in this experiment are weak acids, which undergo dissociation reactions of the type shown in Equation 19.1. The corresponding equation for K_a for these indicators is shown in Equation 19.2.

$$HIn + H_2O \rightleftharpoons H_3O^+ + In^- \quad (19.1)$$

and

$$K_a = \frac{[H_3O^+][In^-]}{[HIn]} \quad (19.2)$$

Because HIn and In^- have different absorptivities at a selected wavelength, absorption spectrophotometry may be used to measure the change in concentration with pH of each form of the indicator. From this information, we can calculate the dissociation constant of the indicator.

Assume that the Beer-Lambert Law is obeyed by both HIn and In^-. If that is so, the relation $A = \varepsilon c \ell$ may be employed to study solutions containing HIn or In^- or both. (As noted in Experiment 17, the Beer-Lambert Law relates the absorbance, A, of a solution to the absorbtivity, ε, of a species that is present in solution at some concentration, c, when light of a particular wavelength traverses a path of length, ℓ, through the solution.)

In the course of this experiment, you will study several solutions. Each will contain the same amount of indicator, but will differ from the others in pH. For instance, one solution will contain a concentration, c, of indicator in the presence of strong acid. In this low pH mixture, virtually all of the indicator will be protonated (i.e., will be in the HIn form). The absorbance of this solution is given in Equation 19.3.

$$A_{\text{low pH}} = A_{\text{HIn}} = \varepsilon_{\text{HIn}} c \ell \quad (19.3)$$

A second solution will contain a concentration, c, of indicator in the presence of strong base. In this high pH mixture, virtually all of the indicator will be deprotonated (i.e., will be in the In^- form). The absorbance of this solution is given by Equation 19.4.

$$A_{\text{high pH}} = A_{\text{In}} = \varepsilon_{\text{In}} c \ell \quad (19.4)$$

In a mixture buffered at intermediate pH, substantial amounts of both HIn and In^- may be present. The concentration of HIn and In^- are then equal to $X_{\text{HIn}} c$ and $(1 - X_{\text{HIn}})c$, respectively, where X_{HIn} represents the mole fraction of indicator in the HIn form and $(1 - X_{\text{HIn}})$ represents the mole fraction of indicator in the In^- form. The absorbance of such a mixture will be the sum of the absorbances of HIn and In^-, as shown in Equation 19.5.

$$A_{\text{intermediate pH}} = A_{\text{HIn}} + A_{\text{In}} = \varepsilon_{\text{HIn}} X_{\text{HIn}} c\ell + \varepsilon_{\text{In}} (1 - X_{\text{HIn}}) c\ell \qquad (19.5)$$

The object of this experiment is to evaluate X_{HIn} for a solution of known pH. Once a value of X_{HIn} has been obtained, it is easy to determine the relative concentrations of HIn and In^- and evaluate K_a. A simple rearrangement of Equation 19.5 shows the relationship between the absorbances of the various solutions and the value of X_{HIn}. (See Equations 19.6 and 19.7.)

$$A_{\text{intermediate pH}} = X_{\text{HIn}} [\varepsilon_{\text{HIn}} c\ell] + (1 - X_{\text{HIn}}) [\varepsilon_{\text{In}} c\ell] \qquad (19.6)$$

$$A_{\text{intermediate pH}} = X_{\text{HIn}} A_{\text{low pH}} + (1 - X_{\text{HIn}}) A_{\text{high pH}} \qquad (19.7)$$

Before making any of the absorbance measurements, you must choose an appropriate wavelength to use. Maximum sensitivity is obtained if the wavelength chosen has the greatest difference between A_{HIn} and A_{In}. This optimum wavelength is easily determined by comparing the spectrum of an indicator in acid solution with the corresponding spectrum of the indicator in basic solution. Figure 19.1 shows the spectra of three different indicators that are suitable for this experiment.

Procedure

Procedure in a Nutshell

Determine the absorbance of an indicator under strongly acidic conditions and under strongly basic conditions. Find the absorbances of mixtures containing substantial amounts of both the acid (HIn) and the base (In^-) form of the indicator. Treat the intermediate absorbances as weighted averages of the absorbances of the acid and base forms of the indicator and find the ratio of $[\text{In}^-]$ to $[\text{HIn}]$ in each intermediate mixture. Determine the pH of each mixture and obtain values of K_a based on the pH of each mixture and the ratio of $[\text{In}^-]$ to $[\text{HIn}]$.

Obtain an indicator assignment from your instructor. Refer to Figure 19.1 and choose the optimum wavelength for your absorbance measurements. Clean five 10-mL test tubes, a buret, and a 2-mL pipet. Use the buret and pipet, respectively, to dispense 5.00 mL of your indicator solution and 2.00 mL of water into each of the five test tubes.

Add 1.00 mL of 1 M HCl to one test tube, 1.00 mL of 1 M NaOH to a second tube, and 1.00 mL of a different buffer solution to each of the remaining test tubes. The buffer solutions to be used will be labeled with nominal pH values. However, the actual pH of each indicator/buffer mixture will be slightly higher than that of pure buffer, owing to the dilution of the buffer with indicator and water. If the actual pH of each diluted buffer is not specified, you will need to measure it accurately using a pH meter.

If possible, familiarize yourself with the operation of the spectrophotometer by studying the instruction manual. Review the material on absorption spectrometry in Experiment 17, if necessary. Follow your instructor's directions concerning the use of the spectrophotometer.

Clean and rinse two matched test tubes or spectrophotometer cells. Fill each cell with water and dry the cell on the outside using a soft, absorbent tissue. Inspect each cell to ensure that no dirt, air bubbles, or scratches are in the light path. Insert each cell in turn into the spectrophotometer and observe the deflection of the meter pointer. The cell that has the lower absorbance should be noted and used as your reference. With that reference cell in the cell compartment, close the cover, and set the light control so that the meter reads 100%T or zero absorbance. Replace the reference with the sample cell, close the cover, and read and record its absorbance.

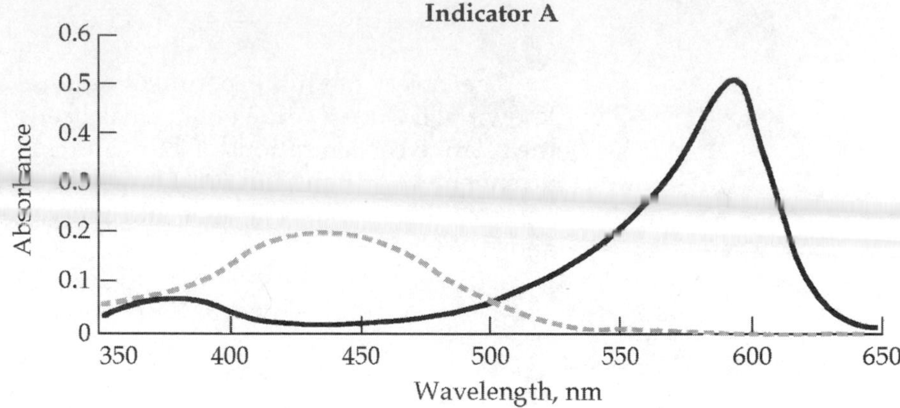

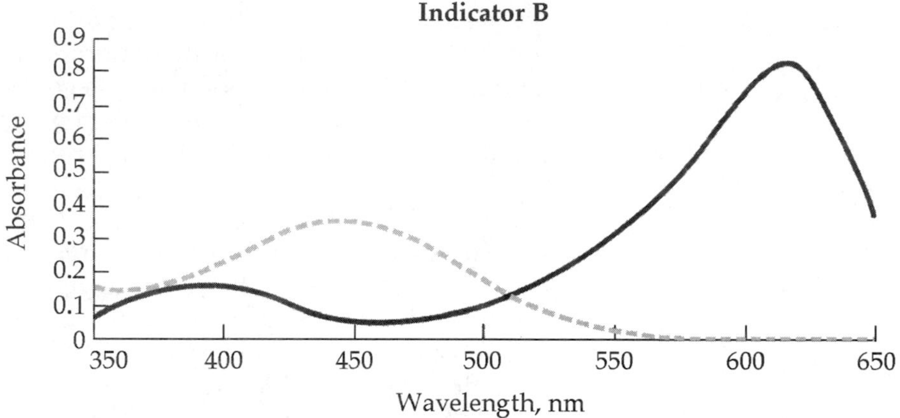

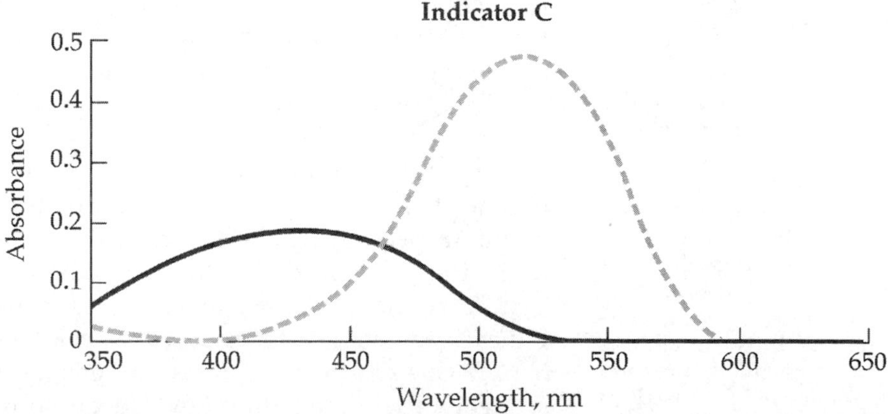

Figure 19.1 Absorption spectrum of an acid-base indicator in 0.125 M HCl (- - -) and in 0.125 M NaOH (——).

Rinse the sample cell with two successive small portions of one of the five indicator solutions prepared previously. Transfer the rest of the solution to the sample cell and insert it into the spectrophotometer. Read and record the absorbance. Repeat these operations for the four remaining indicator solutions. Before reading each sample, check the 0 and 100%*T* settings of the instrument, using your reference cell. Correct the absorbance of each indicator solution for the absorbance of the sample cell.

Using the absorbances of the HCl, the NaOH, and the buffered solutions, calculate the mole fractions of both HIn (X_{HIn}) and In^- (X_{In}) present in each buffered mixture. Report the values of K_a obtained and the average value of K_a. Examine your data carefully before you discard your solutions. If you have not obtained data that will give meaningful values of K_a, obtain or prepare buffers with the pH's you need to make mixtures containing substantial portions of both protonated and deprotonated indicators. Your instructor will give you directions for preparing buffers of intermediate pH from the buffers that have been supplied to you.

Disposal of Reagents

All reagents used in this experiment may be flushed down the drain after they have been neutralized.

Questions

1. If two or more light-absorbing materials are present together, the total absorbance is the sum of the individual absorbances. Using the relation between A and $\%T$, show that the total $\%T$ is not the sum of the individual $\%T$ values.

2. The most accurate results for this experiment are obtained by studying a mixture in which HIn and In^- are present in equal concentration. What is the relation between K_a and $[H^+]$ at this point? Specify the pH of the buffer that should be used to obtain a mixture where $[HIn] = [In^-]$.

3. Is the indicator whose spectrum is shown in Figure 19.2 more highly colored in acid or basic solution? Explain your answer briefly.

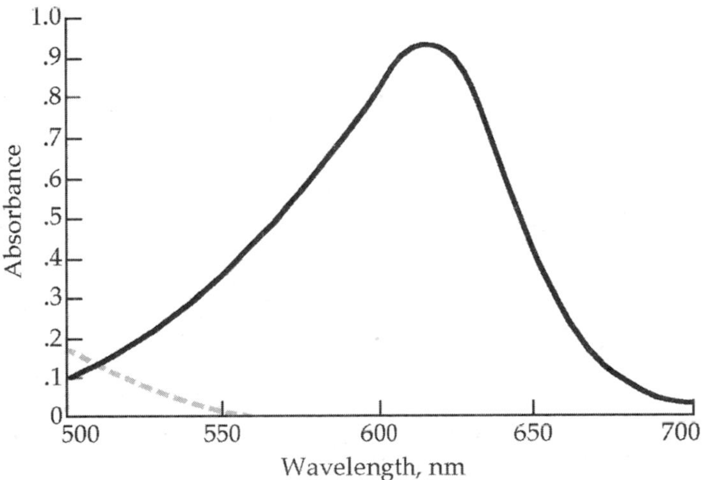

Figure 19.2 Absorption spectrum of an acid-base indicator in 0.1 M HCl (- - -) and in 0.1 M NaOH (——).

name section date

Pre-Lab Exercises for Experiment 19

These exercises are to be completed after you have read the experiment but before you come to the laboratory to perform it.

1. A certain indicator has an absorbance of 0.727 in 1 M NaOH and an absorbance of 0.138 in 1 M HCl. In a solution buffered to a pH of 4.15, the indicator has an absorbance of 0.548. Assume the path length and total concentration of indicator are the same for all mixtures studied and find the value of K_a for the indicator.

2. Jan Student found that the indicator had an absorbance of 0.214 when 5.00 mL of indicator had been mixed with 2.00 mL of water and 1.00 mL of 1 M HCl. Jan also found that the indicator had an absorbance of 0.220 when 5.00 mL of indicator had been mixed with 2.00 mL of water and 1.00 mL of pH 3 buffer (pH of diluted buffer, 3.15). Explain briefly why Jan's results cannot be used to find the value of K_a for the indicator. Give a possible reason for Jan's anomalous results.

name ______ section ______ date ______

Summary Report on Experiment 19

Indicator studied ______

Wavelength used ______

Absorbance of sample cell versus reference cell ______

Solution Number	*Actual pH*	*Absorbance*	*Corrected Absorbance*
1	______	______	______
2	______	______	______
3	______	______	______
4	______	______	______
5	______	______	______
6*	______	______	______
7*	______	______	______
8*	______	______	______

Number	$[H_3O^+]$	X_{HIn}	X_{In}	K_a
3	______	______	______	______
4	______	______	______	______
5	______	______	______	______
6*	______	______	______	______
7*	______	______	______	______
8*	______	______	______	______
			$K_{a,\,av}$	______

* Optional—mixtures to be studied if the others do not give satisfactory results.

Acid Dissociation Constant, K_a [1]

EXP
19P

Objective

In this experiment you will:

- Gain experience mixing solutions of specified concentration.
- Experimentally determine the dissociation constant, K_a, of an acid.
- Investigate the effect of initial solution concentration on the equilibrium constant.

The acid to be used is acetic acid, $HC_2H_3O_2$, and its dissociation equation is:

$$HC_2H_3O_2(aq) \rightleftharpoons H^+(aq) + C_2H_3O_2^-(aq)$$

Materials

CBL 2 interface
TI Graphing Calculator
DataMate program
pH sensor
100-mL beaker
2.00 M $HC_2H_3O_2$
Wash Bottle
Distilled water
100-mL volumetric flask
Pipets
Pipet bulb

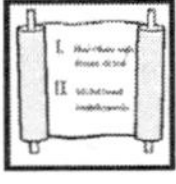

Pre-Lab

1. Write the equilibrium constant expression, K_a, for the dissociation of acetic acid, $HC_2H_3O_2$. (Use Box 3 in the Data and Calculations table of this experiment.)

2. You will be assigned two different $HC_2H_3O_2$ solution concentrations by your TA. Determine the volume, in mL, of 2.00 M $HC_2H_3O_2$ required to prepare each. (Show your calculations and answers in Box 4 of the Data and Calculations table.)

[1] Adapted from Holmquist, D.D., J. Randall, D.L. Volz, *Chemistry with Calculators: Chemistry Experiments Using Vernier Sensors with Texas Instruments CBL 2™* (Vernier Software and Technology, Beaverton, Oregon: 2000). Used with permission.

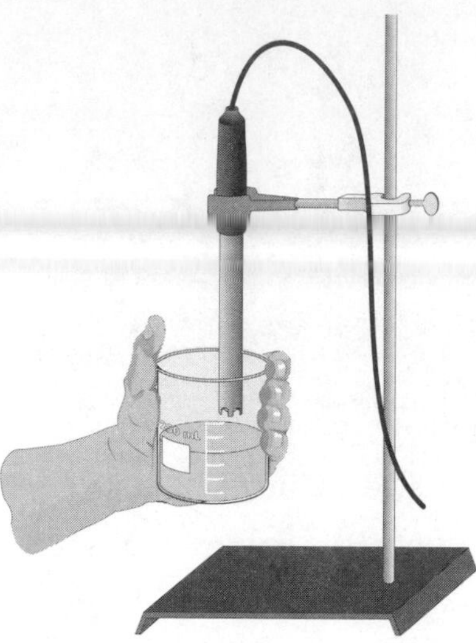

Figure 19P.1

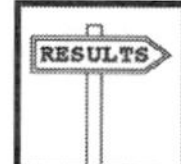

Procedure

1. Obtain and wear safety goggles.

2. Plug the pH sensor into Channel 1 of the CBL 2 interface. Use the link cable to connect the TI Graphing Calculator to the interface. Firmly press in the cable ends.

3. Turn on the calculator and start the DataMate program. Press CLEAR to reset the program.

4. Set up the calculator and interface for the pH sensor.

   a. If the calculator displays pH in CH 1, proceed directly to Step 5. If it does not, continue with this step to set up your sensor manually.

   b. Select SETUP from the main screen.

   c. Press ENTER to select CH 1.

   d. Select PH from the SELECT SENSOR menu.

   e. Select OK to return to the main screen.

5. Put approximately 50 mL of distilled water into a 100-mL volumetric flask.

6. Use a pipet bulb (or pipet pump) to pipet the required volume of 2.00 M acetic acid (calculated in Pre-Lab Step 2) into the volumetric flask. **CAUTION:** *Use care when handling the acetic acid. It can cause painful burns if it comes in contact with your skin or gets into your eyes.* Fill the flask with distilled water to the 100-mL mark. To prevent overshooting the mark, use a wash bottle filled with distilled water for the last few mL. Mix thoroughly.

7. Use a utility clamp to secure the pH sensor to a ringstand as shown in Figure 19P.1.

8. Determine the pH of your solution as follows:

   a. Use about 40 mL of distilled water in a 100-mL beaker to rinse the electrode.

b. Pour about 30 mL of your solution into a clean 100-mL beaker and use it to thoroughly rinse the electrode.

c. Repeat the previous step by rinsing with a second 30-mL portion of your solution.

d. Use the remaining 40-mL portion to determine pH. Swirl the solution vigorously. **Note:** Readings may drift without proper swirling! When the pH reading displayed on the main screen of the calculator stabilizes, record the pH value in your data table. (Round to the nearest 0.01 pH unit.)

e. When done, place the pH sensor in distilled water.

f. Discard the acetic acid solution as directed by your teacher.

9. Repeat Steps 5-8 for your second assigned solution.

10. When you are done, rinse the probe with distilled water and return it to the sensor soaking solution. Select QUIT and exit the DataMate program.

Processing the Data

1. Use the TI calculator to determine the $[H^+]_{eq}$ from the pH values for each solution.

2. Use the obtained value for $[H^+]_{eq}$ and the equation:

$$HC_2H_3O_2(aq) \rightarrow H^+(aq) + C_2H_3O_2^-(aq)$$

to determine $[C_2H_3O_2^-]_{eq}$ and $[HC_2H_3O_2]_{eq}$.

3. Substitute these calculated concentrations into the K_a expression you wrote in Step 1 of the Pre-Lab.

4. Compare your results with those of other students. What effect does initial $HC_2H_3O_2$ concentration seem to have on K_a?

Observations

Data Table

1. Assigned concentration	M	M
2. Measured pH		
3. K_a expression		
4. Volume of 2 M acetic acid	mL	mL
5. $[H^+]_{eq}$	M	M
6. $[C_2H_3O_2^-]_{eq}$	M	M
7. $[HC_2H_3O_2]_{eq}$	M	M
8. K_a calculation		

EXP
20

Temperature Change and Equilibrium

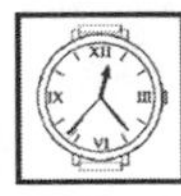

Laboratory Time Required

Three hours.

Special Equipment and Supplies

Buret
Buret Clamp
125-mL Erlenmyer flasks
Pipet, 5 mL
Pipet bulb
Thermometer
Hot plate
100-mL Graduated Cylinder
Borax, $Na_2B_4O_7 \cdot 10H_2O$
0.5 M Hydrochloric acid, HCl, standard solution
Bromcresol green indicator solution
Glass wool

Objective

In performing this experiment, the student will see how the variation of the value of an equilibrium constant with temperature can be used to determine the enthalpy and entropy changes associated with the system in question.

Safety

The dilute HCl and mildly basic borax present no serious health hazard but still should be handled with caution to protect the eyes and skin from possible injury. Safety glasses are required.

Burns or electrical shock may be caused by electrical heating devices that are poorly maintained or carelessly used. Avoid electrical shock by using care when plugging in the power cord and by not using instruments that have frayed cords or that are wet from spilled liquids.

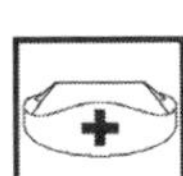

First Aid

Rinse the exposed area thoroughly to flush away chemicals. A doctor should be consulted if discomfort or signs of irritation appear.

In case of electric shock, separate the victim from the source of electricity by turning the power off at the switch box, removing the power cord with a nonconducting tool, or by other means. If the victim is not breathing, CPR should be administered quickly by a qualified person. Get professional help *immediately.*

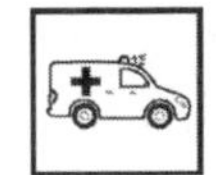

Preamble

Our experience indicates that most physical and chemical changes are affected by temperature. More than a century ago, observations of temperature effects on everyday phenomena led to a formulation of the laws of thermodynamics, which are mathematical descriptions of these observations. The first law tells us about energy; the second, about reversibility and the direction of spontaneous change; and the third, about absolute zero.

The laws of classical thermodynamics apply specifically to the equilibrium condition of a system. When these laws are applied to chemical reactions, they relate the equilibrium condition and the temperature of a given reaction to the thermodynamic quantities, $\Delta H°$, $\Delta S°$, and $\Delta G°$, which specify respectively the change in enthalpy, entropy, and free energy for the system under standard conditions. By studying the effect of temperature on the equilibrium condition for a chemical reaction, we should obtain the fundamental information that we need to evaluate these important thermodynamic functions.

In this experiment, you will investigate the effect of temperature on the solubility of a salt, sodium tetraborate. The data collected will permit the calculation of the equilibrium constant, K_{sp}, associated with the preparation of saturated solutions of the salt. Then you will evaluate $\Delta G°$, $\Delta H°$, and $\Delta S°$ for the dissolution process via graphical procedures.

Principles

The Equilibrium Constant

Many chemical reactions are reversible. At equilibrium, the rate of the forward reaction, in which reactants are converted to products, is equal to the rate of the reverse reaction, in which products of the forward reaction are converted back to the original reactants. When equilibrium is achieved, there is no net change in the relative amounts of reactants and products in the system. These amounts are determined by the value of the equilibrium constant, K, which is defined as shown in Equation 20.1.

$$K = \frac{f(\text{products})}{f(\text{reactants})} \tag{20.1}$$

The nature of the reactants and products determines the function used in the equilibrium expression. Gases are represented as partial pressures (measured in atmospheres); solutes are represented by concentrations (measured in terms of molarity). Pure liquids, pure solids, and solvents are represented by unity.

Thus, the equilibrium between solid calcium carbonate, hydrochloric acid, aqueous calcium chloride, liquid water, and gaseous carbon dioxide would be governed by the K expression shown in Equation 20.2.

$$CaCO_3(s) + 2HCl(aq) \rightleftharpoons CaCl_2(aq) + H_2O(l) + CO_2(g) \tag{20.2}$$

$$K = \frac{[CaCl_2]P_{CO_2}}{[HCl]^2}$$

Variation of K *with Temperature*

Although K is designated an equilibrium "constant," values of K are usually dependent on temperature. If our solutions are ideal and if $\Delta H°$ for the reaction is reasonably constant, we can calculate important thermodynamic quantities from the dependence of K on temperature. We begin with the fundamental expressions:

$$\Delta G^\circ = \Delta H^\circ - T\Delta S^\circ \tag{20.3}$$

$$\Delta G^\circ = -RT\ln K \tag{20.4}$$

These may be combined and rearranged as follows:

$$-RT\ln K = \Delta H^\circ - T\Delta S^\circ \tag{20.5}$$

$$\ln K = -\left(\Delta H^\circ/RT\right) + \left(\Delta S^\circ/R\right) \tag{20.6}$$

Note that Equation 20.6 is of the form

$$y = mx + b \tag{20.7}$$

where $y = \ln K$

$x = 1/T$

$m = -\Delta H^\circ/R$

and $b = \Delta S^\circ/R$

A plot of $\ln K$ versus $1/T$ should therefore be linear, with a slope equal to $-\Delta H^\circ/R$ and an intercept of $\Delta S^\circ/R$.

Solubility of Borax and K_{sp}

The salt chosen for this experiment is sodium tetraborate decahydrate, better known as borax. Borax belongs to a class of compounds, the borates, which contain polyanions composed of trigonal BO_3 and/or tetrahedral BO_4 units linked by bridging oxygen atoms to/from chain or ring structures. The structure of the tetraborate anion, shown in Figure 20.1, illustrates some of these structural features and suggests that it would be more appropriate to represent borax by the formula $Na_2[B_4O_5(OH)_4]\cdot 8H_2O$.

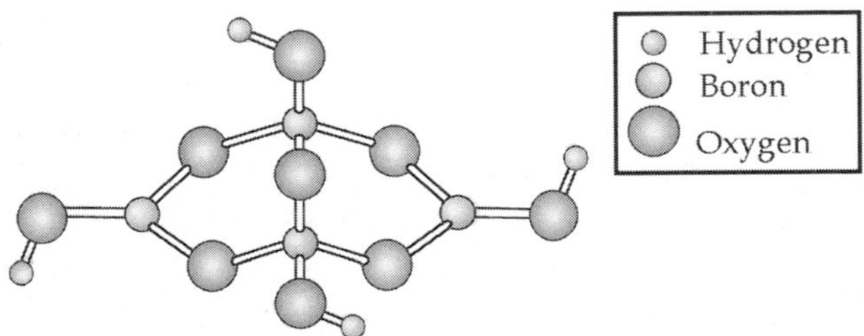

Figure 20.1 Structure of the tetraborate anion

Borax occurs naturally in dry lake beds in California and the southwestern United States, which have long been important sources of this valuable mineral. An early use of borax was in soap and other cleaning products. These products continue to be used today. Borax serves as a flux for solder and is also used in the manufacture of glass.

Borax is mildly basic, as shown in Equation 20.8.

$$B_4O_5(OH)_4^{2-} + 5H_2O \rightleftharpoons 4H_3BO_3 + 2OH^- \qquad (20.8)$$

This property provides the means to determine the solubility of borax. We simply titrate a known volume of the saturated borax solution with a standardized acid solution and calculate the concentration of the tetraborate anions. The reaction of interest is shown in Equation 20.9.

$$B_4O_5(OH)_4^{2-} + 3H_2O + 2H^+ \rightarrow 4H_3BO_3 \qquad (20.9)$$

The formation of a saturated borax solution is represented by Equation 20.10.

$$Na_2[B_4O_5(OH)_4]\cdot 8H_2O(s) \rightleftharpoons 2Na^+(aq) + B_4O_5(OH)_4^{2-}(aq) \qquad (20.10)$$

The equilibrium law expression is therefore:

$$K_{sp} = [Na^+]^2[B_4O_5(OH)_4^{2-}] \qquad (20.11)$$

If we let s represent the concentration of tetraborate ions found by titration with acid, then Equation 20.10 shows that the concentration of sodium ions must be equal to $2s$. Thus, $K_{sp} = 4s^3$.

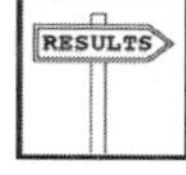

Procedure

Procedure in a Nutshell

Prepare saturated solutions of borax at several temperatures. Determine the value of K_{sp} at these temperatures by titrating the solutions with standard hydrochloric acid. Plot $\ln K$ versus $1/T$ to determine ΔH°_{vap} and ΔS°_{vap} for the reaction:

$$Na_2[B_4O_5(OH)_2]\cdot 8H_2O\,(s) \rightleftharpoons 2Na^+(aq) + B_4O_5(OH)_4^{2-}(aq)$$

Solubility of Borax and K_{sp}

Label five clean 125-mL flasks as follows:

50°C 45°C 40°C 35°C 30°C

Add approximately 5 mL of water to each flask and set the flasks aside.

Weigh approximately 30 to 32 grams of borax and transfer it to a 100-mL beaker; add 50 mL of water (measured in a graduated cylinder) to the borax. Heat the borax/water mixture on a hot plate, stirring occasionally and gently with a thermometer. If the beaker contains undissolved borax when the temperature has reached 65°C, remove the beaker from the hot plate and allow it to cool. If all of the borax dissolves before the mixture reaches 65°C, add more borax to the beaker as it remains on the hot plate. Repeat this process until some borax remains undissolved at 65°C.

Allow the saturated borax solution (i.e., the solution that is in contact with undissolved borax), to cool while stirring and monitoring its temperature carefully. Borax will crystallize from the solution as it cools. Make sure that you are measuring the temperature of the saturated solution and not that of the solid borax that has settled on the bottom of the beaker.

As the solution reaches each of the temperatures marked on the set of flasks containing the 5-mL portions of water, use a pipet and bulb to remove 5.00-mL portions of solution and quickly dispense the saturated solution into the appropriate beaker. (Be careful to avoid drawing borax crystals into the pipet along with the samples of saturated solution.) Record the actual temperature at which each sample of borax was withdrawn.

After pipetting each sample of saturated solution, rinse the pipet with warm distilled water and add the rinsing to the appropriate flask. This introduces a small pipetting error, but it tends to prevent crystallization of the borax in the pipet, which would lead to a much more serious error. When the last sample has been transferred to its flask, add 20 mL of distilled water and 3 to 4 drops of bromcresol green to each sample of the various saturated borax solutions. (Each solution should have a blue tint in the presence of the indicator.)

Rinse and fill a buret with 0.5 M standard hydrochloric acid solution. Record the actual molarity of the HCl solution and the initial buret reading on the Summary Report Sheet. Titrate each blue (saturated borax solution) with HCl, taking as the endpoint a change in indicator color from light blue to pale yellow. Placing a sheet of white paper under the titration vessel will help you to identify the endpoint. Titrate slowly to avoid overshooting the endpoint. Record the final buret reading.

For each temperature studied, calculate the values of K_{sp}, $\ln K_{sp}$, T, and $1/T$. Plot ln K_{sp} versus $1/T$ as described in the Principles section and determine the values of ΔH° and ΔS°. Finally, calculate the value of ΔG° for the dissolution of borax at 298 K.

Disposal of Reagents

The titration solutions may be diluted and then rinsed down the drain. Any remaining saturated solutions containing excess borax can be discarded unless the instructor wishes to recover the borax. Because the solubility of borax decreases rapidly with falling temperatures, such a recovery is easily accomplished. Simply combine the class' solutions, cool the mixture in an ice bath, and collect the borax precipitate on filter paper. Allow the recovered borax to dry in air.

Questions

1. Comment on the significance of each of the possible sources of error listed.

- **a.** The saturated solutions of borax may not show ideal behavior.
- **b.** ΔH° and ΔS° may not be independent of temperature.
- **c.** Obtaining the intercept requires significant extrapolation.

2. Starting with Equation 20.6, show how to arrive at the Clausius-Clapeyron equation:

$$\ln\frac{K_2}{K_1} = \left(\frac{-\Delta H}{R}\right)\left(\frac{1}{T_2} - \frac{1}{T_1}\right)$$

name section date

Pre-Lab Exercises for Experiment 20

These exercises are to be performed after you have read the experiment but before you come to the laboratory to perform it.

1. The solubility of AgCl in water is 3.97×10^{-6} M at 0°C and 1.91×10^{-4} M at 100°C. Calculate K_{sp} for AgCl at each temperature.

2. Calculate $\Delta H°$ for the formation of a saturated solution of AgCl.

3. Calculate $\Delta S°$ for the formation of a saturated solution of AgCl.

4. Calculate $\Delta G°$ for the formation of a saturated solution of AgCl at 25°C.

5. Calculate the slope of a line for which $x_1 = 1.50$, $y_1 = 3.00$, $x_2 = 1.00$, $y_2 = 3.50$. Which graph — (a), (b), or (c) — corresponds to the data given?

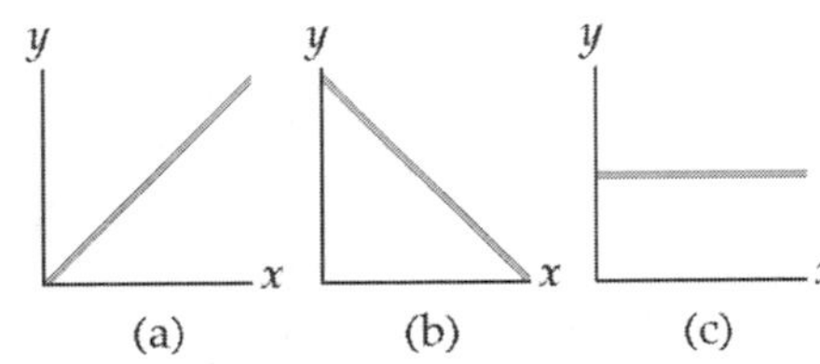

name ____________________ section ____________________ date __________

Summary Report on Experiment 20

Flask label	50°C	45°C	40°C	35°C	30°C
Temperature (°C)					
Temperature (K)					
$1/T$ (K^{-1})					
Actual concentration of HCl solution					
Final buret reading					
Initial buret reading					
Volume of HCl used					
Moles of H^+ used					
Moles of $B_4O_5(OH)_4^{2-}$ reacted					
M, $B_4O_5(OH)_4^{2-}$					
K_{sp}					
$\ln K_{sp}$					

Procedure for calculating $\Delta H°$ and $\Delta S°$ from graphical data:

Results:

$\Delta H°$ ____________________

$\Delta S°$ ____________________

$\Delta G°_{298}$ ____________________

Iodometric Analyses

EXP 21

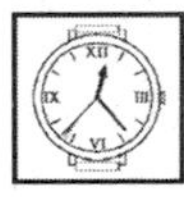

Laboratory Time Required Three Hours.

Special Equipment and Supplies

Option A

Equipment	Supplies
Balance	Bleach, commercial product containing NaOCl, sodium hypochlorite
Burets	Sodium thiosulfate pentahydrate, $Na_2S_2O_3 \cdot 5H_2O(s)$
Buret clamp	Standard 0.02 M potassium iodate, $KIO_3(aq)$
Pipet	Potassium iodide, $KI(s)$
Pipet bulb	0.5 M Sulfuric acid, $H_2SO_4(aq)$
Volumetric flask	1 M Sodium hydroxide, $NaOH(aq)$
	0.2% starch

Option B

Equipment	Supplies
Balance	Pineapple, grapefruit, or white grape juice
Burets	Sodium thiosulfate pentahydrate, $Na_2S_2O_3 \cdot 5H_2O(s)$
Buret clamp	Potassium iodide, $KI(s)$
	Standard 0.02 M potassium iodate, $KIO_3(aq)$
	0.5 M sulfuric acid, $H_2SO_4(aq)$
	1 M sodium hydroxide, $NaOH(aq)$
	0.2% starch

Objective

In performing this experiment, the student will employ iodometric analysis in the determination of the percentage of sodium hypochlorite in household bleach (Option A) or in the determination of the concentration of ascorbic acid in a citrus juice (Option B).

Safety

Either Option Avoid skin contact with potassium iodate, a strong oxidizer, and sulfuric acid, a corrosive acid that is also a good oxidizing agent.

Option A Most household bleaches are alkaline solutions that contain a strong oxidizing agent, the hypochlorite ion. This combination of caustic and oxidizing properties makes bleach solutions hazardous to the eyes and skin.

Option A Exposure of skin to bleach may cause blistering. Serious eye damage may result if bleach gets in the eyes; WEAR GOGGLES WHEN WORKING WITH BLEACH. DO NOT PIPET BY MOUTH. USE THE PIPET BULB.

Option A Chlorine gas might be released when bleach is acidified; chlorine is a poison that can cause lung and eye damage.

Option B Although it may be tempting, do not drink any unused portion of your juice. Drinking is not permitted in the lab. Any unused juice must be discarded.

Safety glasses must be worn; lab coats are advised.

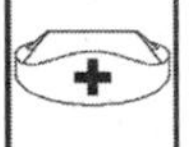

First Aid

Either Option *Skin contact with sodium thiosulfate, potassium iodate, potassium iodide, and starch solutions*: These chemicals present only minor hazards. Flush the chemical from the skin or eyes with copious amounts of water. No additional attention should be needed.

Either Option *Skin or eye contact with bleach, sulfuric acid, or sodium hydroxide*: These solutions present the greatest threat, especially to the eyes. After the chemicals have been thoroughly flushed from the affected areas, consult a doctor immediately if the eyes have been exposed to any of the chemicals or if there is prolonged skin discomfort.

Introduction

Redox titrations, in which one species is oxidized while another is reduced, generally exhibit more complicated stoichiometry than acid/base titrations do. Iodometric analysis is further complicated by the instability of aqueous solutions of iodine and of solutions of sodium thiosulfate (an agent that is generally employed when the chemical amount of iodine in a solution needs to be determined). The two options within this experiment illustrate variations on the theme of iodometric analysis.

Principles

Iodometric Analysis

Aqueous solutions of iodine are notoriously unstable. Despite this, iodometric analysis is a very commonly used laboratory procedure. The solutions may be produced by direct reaction of iodide (I^-) ions with the reagent of interest (hypochlorite (OCl^-) ion in the analysis of bleach (Option A)) or via reaction of iodide ions with iodate (IO_3^-) ions provided by a standard solution of potassium iodate. The latter method is used in the analysis of citrus juices for Vitamin C content. A known amount of iodine is generated, some of it reacts with the Vitamin C, and the amount in excess is determined via titration with thiosulfate ions ($S_2O_3^{2-}$), as shown in Equation 21.1.

$$I_2 + 2S_2O_3^{2-} \rightarrow 2I^- + S_4O_6^{2-} \qquad (21.1)$$

The non-polar iodine ions are not very soluble in water. Therefore, whether the iodine is generated via reaction between potassium iodide and sodium hypochlorite or via reaction between potassium iodide and potassium iodate, an excess of potassium iodide is employed so that the iodine molecules will be transformed into triiodide ions, as shown in Equation 21.2.

$$I_2 + KI \rightarrow KI_3 \quad \text{or} \quad I_2 + I^- \rightarrow I_3^- \qquad (21.2)$$

The stoichiometry of the reaction between thiosulfate ions and triiodide ions is the same as the stoichiometry of the reaction between iodine molecules and thiosulfate ions. (See Equation 21.3.)

$$I_3^- + 2S_2O_3^{2-} \rightarrow 3I^- + S_4O_6^{2-} \qquad (21.3)$$

Option A
Analysis of Bleach for Hypochlorite Content

The chloralkali industry is based on the electrochemical processing of concentrated solutions of sodium chloride, known as brine. The electrolysis of brine produces sodium hydroxide and chlorine gas. Reaction and dilution of these chemicals produces bleach, dilute solutions of sodium hypochlorite. Bleach is used in the home as a cleanser and disinfectant. It is also used in the textile and paper industries to remove colored impurities from cloth and paper.

The active ingredient in bleach is sodium hypochlorite, NaOCl, which is formed by the reaction shown in Equation 21.4. Because this reaction is reversed by acid, bleach solutions must be disposed of properly to avoid the formation of chlorine gas.

$$Cl_2(g) + 2Na^+(aq) + 2OH^-(aq) \rightarrow 2Na^+(aq) + Cl^-(aq) + OCl^-(aq) + H_2O \qquad (21.4)$$

The reactivity of bleach arises from the presence of chlorine in the +1 oxidation state in the OCl^- ion. Because chlorine is a fairly active nonmetal, it is relatively easy to reduce chlorine from the +1 to the –1 state. This makes the hypochlorite ion a good oxidizing agent.

In the analysis of bleach for hypochlorite content that you will perform, a diluted sample of bleach will be allowed to react with excess potassium iodide. In the course of this reaction, the iodide ion will be oxidized to iodine (in the form of the triiodide ion) while the hypochlorite ion will be reduced to chloride. (See Equation 21.5.)

$$OCl^- + 3I^- + 2H^+ \rightarrow I_3^- + Cl^- + H_2O \qquad (21.5)$$

The analysis of bleach for hypochlorite content will be accomplished by determining how much triiodide ion is produced. This is a routine procedure in which a solution of triiodide ion is titrated with a solution of sodium thiosulfate (see Equation 21.3 and Standardization of Thiosulfate Solutions, below).

Option B
Analysis of Juice for Vitamin C Content

Ascorbic acid (Vitamin C) is readily oxidized to dehydroascorbic acid (see Equation 21.6). In this experiment, iodine, in the form of the triiodide complex ion, will be used as the oxidizing agent.

$$C_6H_8O_6 + 3I^- \rightarrow C_6H_6O_6 + I_3^- + 2H^+ \qquad (21.6)$$

This triiodide ion will be generated by the addition of an excess of potassium iodide to a standard solution of potassium iodate. (See Equation 21.7.)

$$IO_3^- + 8I^- + 6H^+ \rightarrow 3I_3^- + 3H_2O \qquad (21.7)$$

The amount of I_3^- generated will be more than sufficient to react with all of the ascorbic acid in the sample. The exact amount of excess triiodide ion will be determined by titration with sodium thiosulfate, $Na_2S_2O_3$ (see Equation 21.3). This method of generating a known amount of reagent, permitting some of it to react with a sample and, ultimately, determining how much of the reagent is left unreacted, is a standard technique, known as a back titration.

Standardization of Thiosulfate Solutions

It is impossible to obtain pure sodium thiosulfate. The decomposition of this unstable solid proceeds even more quickly in solution. Thus, it is necessary, when thiosulfate solutions are needed, to prepare them on the day of their use and to standardize them that day, via titration. The titration process is the same for the standardization of the thiosulfate solution and for the analysis of the iodine content of reaction mixtures.

The reaction between triiodide and thiosulfate ions is followed visually, with the deep brown color of the triiodide ions fading to yellow as the reaction with thiosulfate converts I_3^- ions to the colorless iodide ions. Starch is added shortly before the end point is reached. The remaining triiodide ions form a blue complex with the starch. This complex is then destroyed by the further addition of $S_2O_3^-$ ions. When all I_3^- ions have been converted to I^-, the titration mixture becomes colorless, signaling the end point. (In Option B, the color of the titration mixture will be the underlying color of the juice that is being analyzed by back titration.)

Procedure

Procedure in a Nutshell

Option A Prepare and standardize a solution containing thiosulfate ions. Dilute 10.00 mL of bleach to 100.00 mL. Treat the diluted bleach with an excess of acid and iodide ions, converting the bleach's hypochlorite ions to triiodide ions. Titrate the triiodide ions with the thiosulfate solution and determine the concentration of hypochlorite in both the diluted and undiluted bleach.

Option B Prepare and standardize a solution of sodium thiosulfate. Treat a known volume of standard potassium iodate solution with an excess of acid and potassium iodate, generating a known amount of triiodide ion. Allow some of the triiodide ions to react with the Vitamin C in a known volume of juice. Determine the amount of triiodide *not* reacted with the juice via titration with the standardized thiosulfate solution.

Either Option
Standardization of Thiosulfate Solution

Prepare an approximately 0.08 M thiosulfate solution by dissolving 4 g of $Na_2S_2O_3 \cdot 5H_2O$ in 200 mL of distilled water. Be sure the solution is well mixed. Rinse a buret with two small volumes of distilled water, followed by two small volumes of your thiosulfate solution. Then fill the buret with the solution. Record the initial buret reading.

Record the molarity of the standard potassium iodate solution provided for your use. Rinse a second buret with two small volumes of distilled water, followed by two small volumes of the standard KIO_3 solution. Then fill the second buret with the iodate solution and record the initial buret reading for the second buret. Deliver 20 mL of KIO_3 solution into a clean, 250-mL Erlenmeyer flask and record the final buret reading for the iodate buret. Add 20 mL of distilled water and 2 g of potassium iodide to the flask. Swirl the flask to dissolve the solid potassium iodide. Continue to swirl the flask while adding 20 mL of 0.5 M H_2SO_4. The mixture in the Erlenmeyer flask should take on the deep red-brown color characteristic of I_3^- ions.

Titrate the I_3^- ions with $S_2O_3^{2-}$, adding 1-mL increments of thiosulfate initially, but reducing the increment size as the color of the titration mixture turns from brown to yellow. Then add 5 mL of 0.2% starch solution. Consult your instructor if the titration mixture does not develop a blue color after starch has been added and the flask has been swirled. Use your wash bottle to rinse all splattered drops from the walls of the flask into the titration mixture. Resume the addition of thiosulfate, adding titrant by drops until the blue color of the starch–iodine complex disappears completely. Usually, only a few additional drops are required. Record the final reading of the thiosulfate buret and calculate the molarity of your thiosulfate solution. Repeat the procedure. Report the molarity obtained in each trial and the average molarity.

Option A
Titration of Bleach

Refill your buret with $Na_2S_2O_3$ solution and record the initial reading. Clean a 10-mL volumetric pipet thoroughly. Rinse it twice with distilled water, followed by two small volumes of household bleach. Then use the pipet to deliver 10.00 mL of bleach to a clean (but not necessarily dry) 100-mL volumetric flask.

Caution!

AVOID GETTING BLEACH IN YOUR EYES. IF BLEACH DOES GET IN YOUR EYES, RINSE THEM QUICKLY AND THOROUGHLY (AT LEAST FIFTEEN MINUTES) AT THE EYEWASH FOUNTAIN. SEEK MEDICAL ATTENTION.

Add 50 mL of distilled water to the bleach and swirl the flask carefully to mix its contents. Dilute the solution with distilled water to the 100-mL mark, then invert the flask several times to promote mixing of the solution.

Dissolve approximately 2 g of KI in a mixture of 20 mL of distilled water and 20 mL of 0.5 M H_2SO_4 in a 250-mL Erlenmeyer flask. Swirl the flask until all solid has dissolved. If the solution turns yellow at this point, discard it. Prepare a new solution, being sure that the flask is clean and that the H_2SO_4 solution is at or below room temperature.

Rinse the pipet with two small portions of distilled water, followed by two small portions of the diluted bleach. Use the rinsed pipet to deliver 10.00 mL of diluted bleach to the flask containing the solution of KI. Swirl the Erlenmeyer flask to allow the contents to mix. Using the procedure for the standardization of the thiosulfate solution, titrate the triiodide ions that result from the reaction of the bleach and the KI solution. Repeat the titration. Report the experimental molarity of the diluted bleach for each titration, and the average molarity. Also report the calculated molarity of commercial bleach.

Option B
Titration of a Sample of Juice

Use a graduated cylinder to measure 20 mL of pineapple, grapefruit, or white grape juice. Transfer the juice to a clean 250-mL Erlenmeyer flask. Add 20 mL of 0.5 M H_2SO_4 to the flask containing the juice. Swirl the flask.

Fill the burets once again with the appropriate solutions. Add 2 g of potassium iodide and 20 mL of potassium iodate to the Erlenmeyer flask containing the juice and sulfuric acid. Titrate immediately with sodium thiosulfate. Add starch, as before, when the iodine color fades. Continue titrating until the end point has been reached.

Repeat the titration. Report the chemical amount of Vitamin C found in each portion of the juice and the molarity of Vitamin C in each sample. Determine the number of mg of Vitamin C contained in a 6-oz. serving of juice.

Disposal of Reagents

Option A The potassium iodate solution should be reduced to potassium iodide before disposal. This may be achieved by following essentially the same procedure as used in the experiment. Dissolve excess KI in the KIO_3 solution. Then add a few mL of 0.5 M H_2SO_4 to acidify the mixture. Mix well. Next add sodium thiosulfate solution until the yellow-brown iodine color has disappeared. Dilute the solution and rinse it down the drain.

The 0.5 M H_2SO_4 solution should be neutralized with aqueous NaOH, and then rinsed down the drain.

All other solutions may be diluted and then poured down the drain with a steady stream of water. Do not dispose of undiluted bleach as it may react with other chemicals in the drainage system to produce poisonous chlorine gas anywhere in the drain pipes.

Option B The potassium iodate solution should be reduced to potassium iodide before disposal. This may be achieved by following essentially the same procedure used in the experiment. Dissolve excess KI in the KIO_3 solution. Then add a few mL of 0.5 M H_2SO_4 to acidify the mixture. Mix well. Add sodium thiosulfate solution until the yellow-brown iodine color has disappeared. Dilute the solution and flush it down the drain.

The 0.5 M H_2SO_4 solution should be neutralized with aqueous NaOH and then flushed down the drain.

All other solutions may be diluted and flushed down the drain.

Questions

Option A

1. Most commercial bleaches are 5.25% sodium hypochlorite (by mass). Assume the density of such a bleach is 1.02 g/mL and calculate the molarity of the commercial bleach with respect to hypochlorite ion.

2. How does the molarity obtained in answer to Question 1 compare with your experimental value of the molarity?

3. Mixing bleach with hydrochloric acid can result in the release of noxious chlorine gas. For this reason, many household cleansers, which are very acidic, carry warning labels stating that they should not be mixed with bleach. Why were you able to mix samples of bleach with acid in this experiment and not produce chlorine?

Option B

1. A 20.0 mL sample of white grape juice was titrated to the phenolphthalein end point with 13.65 mL of 0.1097 M NaOH. Assume all of the acid in the juice sample was ascorbic acid and calculate the molarity of ascorbic acid in the white grape juice.

2. Calculate the number of milligrams of ascorbic acid per six ounce serving of juice, based on your answer to Question 1. (1 L = 33.6 oz.)

3. Your answer to Question 2 is likely to be in error. Explain why it should be obvious that the answer is not correct and also explain which assumption made in obtaining the answer is not valid. (Hint: Find out why certain fruits are classified as citrus fruits.)

name   section   date

Pre-Lab Exercises for Experiment 21

These exercises are to be completed after you have read the experiment but before you come to the laboratory to perform it.

Option A
Analysis of Bleach for Hypochlorite Content

1. A 10-mL pipet was used to deliver a sample of commercial bleach to a 100-mL volumetric flask. Distilled water was added and the flask was inverted several times to produce a uniform, 100.00-mL sample of diluted bleach. The pipet was rinsed with diluted bleach and then used to deliver 10.00-mL of diluted bleach to an Erlenmeyer flask. Sufficient quantities of H_2SO_4 and KI were added to convert all OCl^- ions to chloride ions, with consequent conversion of some iodide ions to I_3^- ions. The 10.00 mL of diluted bleach required the addition of 24.76 mL of 0.0537 M thiosulfate for complete reaction. Use this data to compute the molarity of the undiluted bleach with respect to sodium hypochlorite.

2. Use your answer to Question 1 to find the percent NaOCl in the undiluted bleach. State any assumptions you make in performing the calculation.

Pre-Lab Exercises for Experiment 21

These exercises are to be completed after you have read the experiment but before you come to the laboratory to perform it.

Option B
Analysis of Juice for Vitamin C Content

1. A 20.0-mL portion of pineapple juice was placed in a flask with 20 mL of 0.5 M H_2SO_4. Then 2 g of potassium iodide and 20.17 mL of 0.02001 M potassium iodate were added to the sample. Titration with sodium thiosulfate required the use of 18.94 mL of 0.07992 M $Na_2S_2O_3$. Find the molarity of ascorbic acid in the juice.

2. Use your answer to Question 1 to determine the number of milligrams of Vitamin C in a 6–oz. serving of the pineapple juice that was studied.

name ______ section ______ date ______

Summary Report on Experiment 21

Option A
Analysis of Bleach for Hypochlorite Content

Molarity of KIO_3 ______

Standardization of Thiosulfate Solution

	Trial 1	*Trial 2*	*Trial 3**
Final buret reading, KIO_3	______	______	______
Initial buret reading, KIO_3	______	______	______
Volume of KIO_3 delivered	______	______	______
Final buret reading, $Na_2S_2O_3$	______	______	______
Initial buret reading, $Na_2S_2O_3$	______	______	______
Volume of $Na_2S_2O_3$ used	______	______	______
Molarity of $Na_2S_2O_3$	______	______	______
Average molarity of $Na_2S_2O_3$		______	

Titration of Hypochlorite

	Trial 1	*Trial 2*	*Trial 3**
Volume of diluted bleach	______	______	______
Final buret reading, $Na_2S_2O_3$	______	______	______
Initial buret reading, $Na_2S_2O_3$	______	______	______
Volume $Na_2S_2O_3$ used	______	______	______
Molarity of NaOCl in dilute bleach	______	______	______
Average molarity of NaOCl in dilute bleach		______	
Molarity of NaOCl in undiluted bleach		______	
Percent NaOCl by mass in undiluted bleach		______	

* Optional

Option B
Analysis of Juice for Vitamin C Content

Standardization of Sodium Thiosulfate

Molarity of KIO_3 ______________

	Trial 1	*Trial 2*	*Trial 3**
Final buret reading, KIO_3	______	______	______
Initial buret reading, KIO_3	______	______	______
Volume of KIO_3 delivered	______	______	______
Final buret reading, $Na_2S_2O_3$	______	______	______
Initial buret reading, $Na_2S_2O_3$	______	______	______
Volume of $Na_2S_2O_3$ used	______	______	______
Calculated molarity of $Na_2S_2O_3$	______	______	______
Average molarity of $Na_2S_2O_3$		______	

Titration of a Sample of Juice

	Trial 1	*Trial 2*	*Trial 3**
Volume of juice used	______	______	______
Final buret reading, KIO_3	______	______	______
Initial buret reading, KIO_3	______	______	______
Volume of KIO_3 delivered	______	______	______
Final buret reading, $Na_2S_2O_3$	______	______	______
Initial buret reading, $Na_2S_2O_3$	______	______	______
Volume of $Na_2S_2O_3$ used	______	______	______
Moles of ascorbic acid reacted	______	______	______
Average mg Vitamin C per 6-oz. juice		______	

* Optional

Electrochemical Cells

EXP
22

Laboratory Time Required

Two hours. May be combined with Experiment 24.

Special Equipment and Supplies

Hexagonal weighing boats
Light cardboard
Double-sided tape
Masking tape
Strips of filter paper (for use as salt bridges)
Electrical leads
pH meter
Sandpaper
Scissors
Forceps
Wooden applicator stick

0.1 M zinc nitrate, $Zn(NO_3)_2(aq)$
0.1 M copper(II) nitrate, $Cu(NO_3)_2(aq)$
0.1 M silver nitrate, $AgNO_3(aq)$
0.1 M lead(II) nitrate, $Pb(NO_3)_2(aq)$
0.1 M tin(II) chloride, $SnCl_2(aq)$
0.1 M potassium nitrate, $KNO_3(aq)$
12 M hydrochloric acid, $HCl(aq)$
1.0 M potassium iodide, $KI(aq)$
3.0 M ammonia, $NH_3(aq)$
Electrodes (strips of Zn, Cu, Ag, Sn, and Pb foils)

Objective

In performing this experiment, the student will construct galvanic cells and determine the value of $\Delta\varepsilon°$ for each. In addition, the student will evaluate a number of standard reduction potentials and evaluate several equilibrium constants by use of electrochemical data.

Safety

Avoid electrical shock when disconnecting or connecting the pH meter power cord.

The Ag^+ and Pb^{2+} ions are **toxic** and their solutions should be handled carefully to avoid skin contact. Wash your hands after sanding and handling the lead electrode.

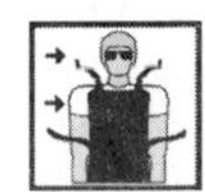

The 12 M hydrochloric acid is corrosive and may cause burns or skin irritation.

Avoid breathing HCl fumes. Pipet the solutions (using pipet and bulb) or pour solutions carefully.

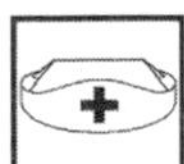

First Aid

Remove spilled silver and lead solutions from the skin by washing thoroughly with water.

Preamble

Electrochemical cells are devices that allow the interconversion of electrical and chemical energy. In voltaic (or galvanic) cells, a spontaneous chemical reaction produces electricity.

Voltaic cells are often used to determine the relative activities of metals. They may also be used to determine equilibrium constants. These applications are explored in this experiment.

Principles

Determination of Standard Reduction Potentials

A typical electrochemical cell, such as the one shown in Figure 22.1, consists of two half-cells linked by a wire and a salt bridge. Each half-cell consists of a metal electrode in contact with a solution containing a salt of that metal. One half-cell functions as the anode, where the oxidation reaction ($M \rightarrow M^{n+} + ne^-$) takes place; the other half-cell functions as the cathode, where the reduction reaction ($M^{n+} + ne^- \rightarrow M$) takes place. Electrons flow from the anode to the cathode via the wire. The salt bridge allows migration of ions to prevent an imbalance of charge from building up as electrons leave the anode and move to the cathode.

Inserting a voltmeter into the circuit between the half-cells permits a measurement of the voltage — or potential difference — between the half-cells. In general, this voltage is designated by the symbol, $\Delta\varepsilon$. When the solutions in the half-cells are 1 M with respect to the ions involved in the oxidation and reduction reactions, the cell is designated a "standard cell" and its voltage is called a "standard potential difference," $\Delta\varepsilon^\circ$. Many textbooks and some reference books, such as the *CRC Handbook of Chemistry and Physics,* contain tables of standard reduction potentials, which show the values of ε° for various reduction reactions of the type $M^{n+} + ne^- \rightarrow M$. The ε°'s in the tables were obtained by assigning a potential of 0.00 V to the reduction reaction: $2H^+ (1\ M) + 2e^- \rightarrow H_2$ (*g*, 1 atm) and measuring the potentials of other standard half-cells coupled to this standard hydrogen electrode.

Whenever two standard half-cells are coupled (joined to create a voltaic cell), the one with the less positive (or more negative) value of ε° functions as the anode. The value of $\Delta\varepsilon^\circ$ for the cell is given by the difference between ε° for the half-cell that functions as the cathode and ε° for the half-cell that functions as the anode. The use of standard reduction potentials to predict the potential of an electrochemical cell is illustrated below. (See Equations 22.1–22.5.)

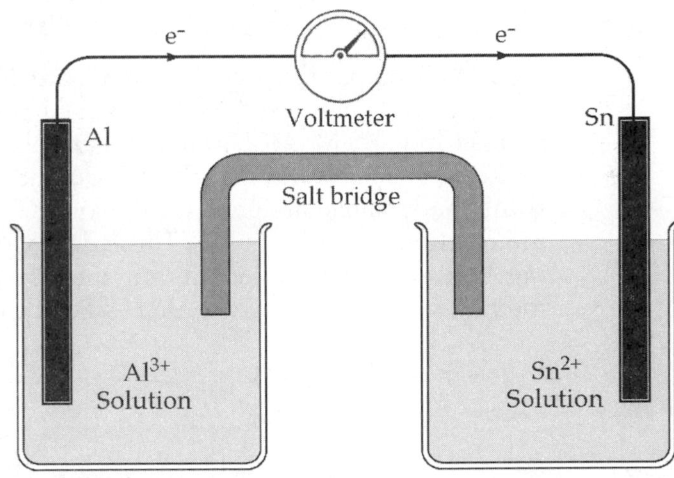

Figure 22.1 An electrochemical cell

	Reduction Reaction	$\varepsilon°$	
	$Sn^{2+} + 2e^- \rightarrow Sn$	–0.14 V	(22.1)
	$Al^{3+} + 3e^- \rightarrow Al$	–1.66 V	(22.2)
Cathode:	$(Sn^{2+} + 2e^- \rightarrow Sn) \times 3$	$\varepsilon°_{cathode} = -0.14$ V	(22.3)
Anode:	$(Al^{3+} + 3e^- \rightarrow Al) \times 2$	$-\varepsilon°_{anode} = -(-1.66$ V$)$	(22.4)
Cell reaction:	$3Sn^{2+} + 2Al \rightarrow 3Sn + 2Al^{3+}$	$\Delta\varepsilon° = 1.52$ V	(22.5)

The electrochemical cells that you will construct in the performance of this experiment will not be standard cells. However, you will be able to use the potentials you observe to obtain standard reduction potentials by the use of the Nernst equation. (See Equation 22.6.)

$$\Delta\varepsilon = \Delta\varepsilon° - \frac{0.0257\ \text{V}}{n}\ln Q \qquad (22.6)$$

To use the Nernst equation, you need to know the value of n, which represents the number of electrons transferred in the cell reaction, and the form of Q, which is called the **reaction quotient.** This quotient is the ratio of the concentrations of ions that are products to the concentrations of ions that are reactants, and may be represented as $[P]^a/[R]^b$, where a and b are the coefficients that appear in the equation for the cell reaction. The number 0.0257 is the numerical value of several constants including the temperature (T = 298 K) and the Faraday (the charge on a mole of electrons); note that this term has the unit of volts. The relationship between Δε for a $Sn/Sn^{2+}(0.10\ \text{M}) - Al/Al^{3+}(0.10\ \text{M})$ cell and $\Delta\varepsilon°$ is shown in Equations 22.7 and 22.8. Note that n = 6, a = 2, and b = 3 for this case.

$$\Delta e = \Delta e° - \frac{0.0257\ \text{V}}{6}\ln\frac{[Al^{3+}]^2}{[Sn^{2+}]^3} \qquad (22.7)$$

$$\Delta\varepsilon = 1.52 \text{ V} - 0.0099 \text{ V} = 1.51 \text{ V} \qquad (22.8)$$

Let us suppose that you prepared an electrochemical cell composed of a tin electrode immersed in 0.01 M $SnCl_2$ and an aluminum electrode immersed in 0.010 M $AlCl_3$. Further suppose you found that the aluminum half cell was the anode and the cell potential was 1.50 V. Let us further assume that you had assigned a value of 1.66 V to the reduction potential of the reaction $Al^{3+} + 3e^- \rightarrow Al$. These data would be sufficient for determining the reduction potential for the reaction $Sn^{2+}(1\text{ M}) + 2e^- \rightarrow Sn$. A method for doing so is illustrated in Equations 22.9–22.11.

$$\Delta\varepsilon = (\varepsilon^\circ_{\text{cathode}} - \varepsilon^\circ_{\text{anode}}) - \frac{0.0257\text{V}}{6})\ln\frac{[Al^{3+}]^2}{[Sn^{2+}]^3} \qquad (22.9)$$

$$1.50 \text{ V} = (\varepsilon^\circ_{\text{cathode}} - 1.66 \text{ V}) - \frac{0.0257 \text{ V}}{6}\ln\frac{(0.010)^2}{(0.010)^3} \qquad (22.10)$$

$$\varepsilon_{\text{cathode}} = -0.14 \text{ V} \qquad (22.11)$$

In this experiment, you will prepare an apparatus that will permit you to couple the half-cells pairwise, so that you will study each of the half-cells in combination with each of the other half-cells. Using the data you obtain and the Nernst equation, you will be able to evaluate ε° for the equations $M^{2+} + 2e^- \rightarrow M$, where M = Cu, Sn, Pb, and Zn, assuming a value of 0.80 V for the reaction $Ag^+ + e^- \rightarrow Ag$. This choice of reference will facilitate comparison of your experimental values of ε°'s to the literature values for these standard reduction potentials.

Evaluating Equilibrium Constants

In addition to determining the values of standard reduction potentials, one can use electrochemical cells to evaluate certain equilibrium constants, as illustrated below.

Consider an electrochemical cell that consists of two identical half-cells, each composed of a silver electrode suspended in a solution containing silver ions at some molarity (Figure 22.2). The voltage of the cell will be zero because each half-cell is identical to the other.

Now imagine that some hydrochloric acid (enough to cause the precipitation of silver chloride) is added to one half-cell. A voltage will now be measured in the electrochemical cell because the concentrations of Ag^+ will differ in the two half-cells (Figure 22.3). Such a cell is called a "concentration cell."

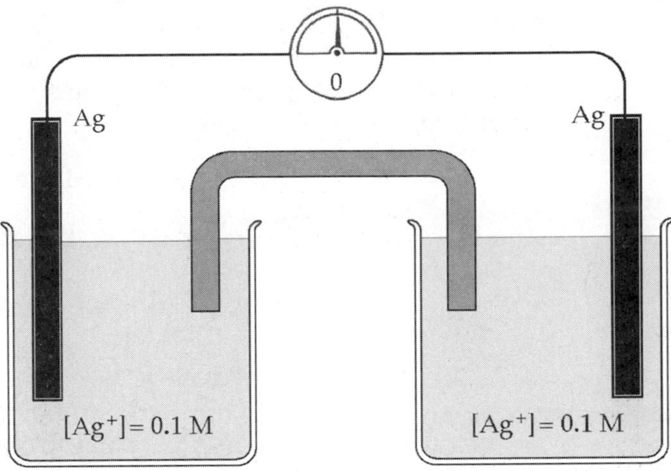

Figure 22.2 Linking two identical half-cells produces no voltage

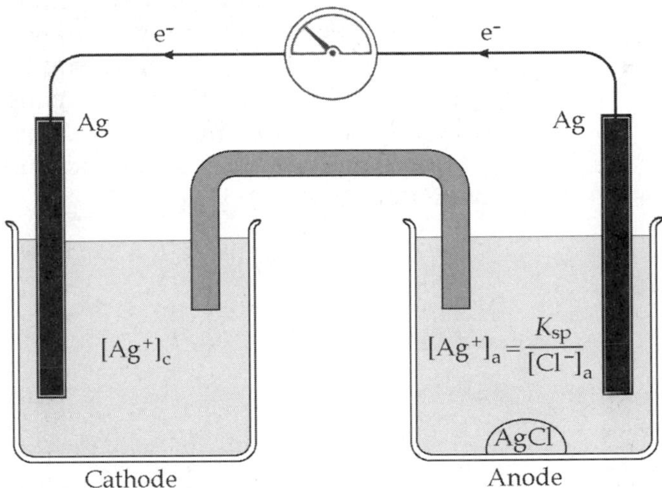

Figure 22.3 Direct coupling of silver/silver nitrate and silver/silver chloride half-cells

In Figure 22.3, the half-cell containing the silver chloride precipitate is designated as the anode, while the half-cell containing 0.10 M silver nitrate is designated as the cathode. Why is this so? A concentration cell will show a non-zero voltage only so long as there is a difference in the concentration of the ions in the two half-cells that comprise the cell. Thus, the spontaneous reaction is the one that tends to equalize the ion concentrations in the two half-cells. The half-cell containing the more concentrated solution becomes the cathode, and the concentration of ions decreases as the reaction $M^{n+} + ne^- \rightarrow M$ occurs. Conversely, the half-cell containing the less concentrated solution becomes the anode, and the concentration of ions increases via the reaction $M \rightarrow M^{n+} + ne^-$. The potential of the cell decreases steadily, finally reaching a value of zero when the concentration of ions in both half-cells becomes the same. In the cell pictured in Figure 22.3, the concentration of silver ions is 0.1 M in each half-cell before the addition of chloride ions. Following the addition of chloride ions to one half-cell, the concentration of silver ions in that half-cell decreases because some silver ions are removed from solution as the precipitate of silver chloride forms. This situation is summarized in Equations 22.12–22.15, where X is the value of the silver ion's concentration in the half-cell containing chloride ions.

Cathode: $$Ag^+(0.10\ M) + e^- \rightarrow Ag \quad (22.12)$$

Anode: $$Ag \rightarrow Ag^+(X\ M) + e^- \quad (22.13)$$

Overall: $$Ag^+(0.10\ M) + Ag \rightarrow Ag + Ag^+\ (X\ M) \quad (22.14)$$

$$\Delta\varepsilon = \Delta\varepsilon^\circ - \frac{0.0257\ V}{1}\ln\frac{X}{0.10} \quad (22.15)$$

The value of X may be obtained readily from an experimentally determined value of $\Delta\varepsilon$, because $\Delta\varepsilon^\circ$ for a concentration cell must equal zero. (Why?) If the cell is constructed with a known excess chloride ion concentration, the value of K_{sp} $(= X[Cl^-])$ can also be obtained.

Arguments similar to those made above can be used to evaluate a number of K_{sp}'s. In this experiment, you may be asked to evaluate K_{sp} for $AgCl$, AgI, and/or PbI_2. Still other equilibrium constants may be evaluated, provided that the corresponding equilibria result in a change in concentration of a metal ion involved in an oxidation/reduction reaction. For instance, the concentration of silver ions may be reduced by complexing Ag^+ with ammonia. (See Equation 22.16.)

$$Ag^+(aq) + 2NH_3(aq) \rightarrow Ag(NH_3)_2^+(aq) \quad (22.16)$$

The form of the equilibrium constant, K_f, for this reaction is shown in Equation 22.17.

$$K_f = \frac{[Ag(NH_3)_2^+]}{[Ag^+][NH_3]^2} \quad (22.17)$$

The value of K_f for $Ag(NH_3)_2^+$ may be obtained by coupling a half-cell containing a mixture of silver nitrate and ammonia to a silver/silver nitrate half-cell. The silver ion concentration, X, is obtained from Equation 22.15. Of course, K_f can be evaluated only if the ammonia is in excess and conditions of mixing are such that $[Ag(NH_3)_2^+]$ and $[NH_3]$ can be evaluated. In this experiment, you may be asked to evaluate K_f for $Ag(NH_3)_2^+$, $Cu(NH_3)_4^{2+}$, and/or $Zn(NH_3)_4^{2+}$.

Although concentration cells are frequently used to evaluate equilibrium constants, it is possible to obtain K values from any voltaic cell in which a half-cell of known potential is coupled to a half-cell in which a chemical reaction limits the concentration of the ion of interest. In this experiment, you will obtain values of K_{sp} for $AgCl$ from a concentration cell and from a cell in which a $AgCl/Ag$ half-cell is coupled to a Zn^{2+}/Zn half-cell.

RESULTS

Procedure

A. Measurement of Standard Reduction Potentials

Crossing three strips of double-sided tape, make an "asterisk" on a cardboard backing. Press a weighing boat firmly onto the middle of the vertical strip of tape. Then press five additional weighing boats onto the cross strips of tape, creating an open honeycomb pattern (Figure 22.4). Place approximately 5 mL of 0.10 M KNO_3 in the center boat. Place approximately 5 mL of 0.10 M $AgNO_3$, 0.10 M $Cu(NO_3)_2$, 0.10 M $Pb(NO_3)_2$, 0.10 M $Zn(NO_3)_2$, and 0.10 M $SnCl_2$ respectively, in each of the outer boats, labeling the boats by showing the symbol for each cation on the cardboard backing to which the weighing boats are attached. Be sure that the contents of the various boats are not allowed to mix.

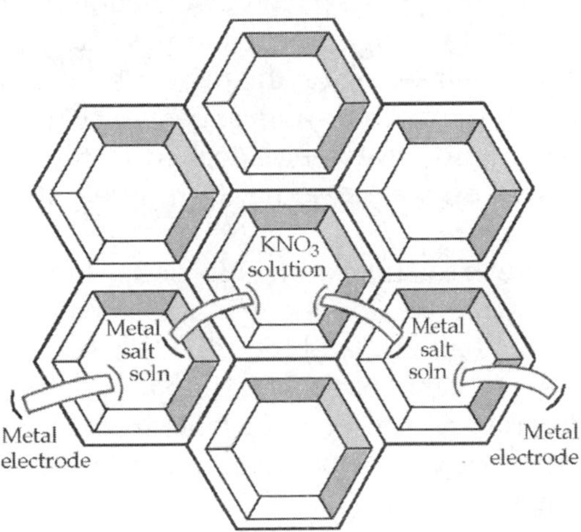

Figure 22.4 Electrochemical cells

Procedure in a Nutshell

Construct electrochemical cells and measure the potential difference ($\Delta\mathcal{E}$) between each pair of half-cells. Use the Nernst equation to convert the observed potential differences to standard potential differences ($\Delta\mathcal{E}°$). Assign the Ag^+/Ag a standard reduction potential ($\mathcal{E}°$) of 0.80 V and find the values of $\mathcal{E}°$ for the Zn^{2+}/Zn, Cu^{2+}/Cu, Sn^{2+}/Sn, and Pb^{2+}/Pb half-cells. Construct electrochemical cells in which a precipitation or complexation reaction has altered the concentration of metal ion in one half-cell. Using the measured value of $\Delta\mathcal{E}$ for such a cell and the Nernst equation, determine the value of $[M^{n+}]$ in the half-cell where the reaction occurred. Use stoichiometric analysis to determine the concentrations of other species involved in the reaction and find the value of the equilibrium constant for the reaction.

Obtain strips of Ag, Cu, Pb, Zn, and Sn foils, which will serve as electrodes. Lift each metal electrode with forceps and sand it to remove any oxide coating. Place one end of the electrode in the solution of the corresponding ion (that is, Zn in Zn^{2+}, etc.) to complete each half-cell. (Do not handle the electrodes with your fingers; use the forceps at all times.) You may bend the electrodes, if necessary. You may also use a small piece of masking tape to secure the electrodes to the weighing boat edge or the cardboard on which the boats are mounted.

Attach the electrical leads to the pH meter. Position the weighing boat honeycomb so that both electrical leads can reach around it easily. Then use masking tape to secure the cardboard holding the honeycomb to the laboratory bench. Plug in the pH meter, with the function selection knob set on standby. Then turn the function selection knob to mV (or +mV) and touch the leads together. Use the calibration knob to set the meter reading to its lowest possible reading.

Dip one end of a strip of filter paper into the solution of zinc nitrate. Place the other end of the filter paper in the solution of potassium nitrate. (You may use a wooden applicator stick, if necessary, to submerge the second end of the paper in the potassium nitrate solution.) The filter paper "bridges" can function effectively as "salt bridges" (conduits for spectator ions) only if the entire length of the filter paper is moistened by the solutions in which the paper is immersed.

Use separate strips of filter paper to link the solutions of copper(II) nitrate, lead nitrate, silver nitrate, and tin(II) chloride to the central boat containing the solution of potassium nitrate. Two strips of filter paper, connecting any two outer boats to the center boat, constitute the salt bridge of your electrochemical cell.

Attach the alligator clip of one of the electrical leads to the zinc electrode. Attach the alligator clip of the other electrical lead to the silver electrode. If the needle of the pH meter goes off scale, indicating a value below –700 mV (or below 0.0 pH), reverse the clips on the electrodes.

Wait 15 seconds, then record the reading on the meter. Use a label (or a piece of masking tape) to mark the electrical lead to which the zinc metal electrode is attached. The lead with the tape will always be the anode in any cell that gives a positive deflection on the pH meter.

Move the alligator clips to new electrodes. Wait 15 seconds before recording the reading on the pH meter. For each cell, note the identity of the electrodes involved, and state which is the anode. Remember that the leads should always be attached to the electrodes in a manner that produces an on-scale reading. Continue until all of the boats have been coupled in pairs.

Convert your readings to volts. Then use the Nernst equation to find the value of $\varepsilon°$ for the Zn^{2+}/Zn half-cell, the Cu^{2+}/Cu half-cell, the Sn^{2+}/Sn half-cell, and the Pb^{2+}/Pb half-cell, assigning the Ag^{+}/Ag half-cell an $\varepsilon°$ value of 0.80 V. Prepare a table of standard half-cell potentials and compare it to one available in the literature. Comment briefly on the accuracy of your experimental work. Briefly explain any error in your experimental results.

B. *Evaluation of Equilibrium Constants*

Insert a sixth weighing boat in the open space of your honeycomb. Fill this cell with 5.0 mL of 0.10 M $AgNO_3$ and 0.50 mL of 12 M HCl. Immerse a freshly sanded piece of silver foil into this mixture and connect the sixth boat to the central boat, as usual, with a strip of filter paper. Attach the alligator clips to the two silver foils and wait 15 seconds before recording the reading on your pH meter. Note which half-cell is the anode.

Move the alligator clip from the silver foil that is immersed in silver nitrate to the zinc electrode. Wait 15 seconds before recording the reading on the pH meter. Note which half-cell is the anode.

Remove the weighing boat that contains the silver chloride from the honeycomb. Clean (see Disposal of Reagents) and dry the boat. Place it in the honeycomb once again and fill it, as your instructor directs, using a foil and mixture of solutions specified in Table 22.1. Use the potential of the resulting concentration cell to find the value of your assigned K.

Disposal of Reagents

The $Zn(NO_3)_2$, $Cu(NO_3)_2$, $SnCl_2$ and KNO_3 solutions may be diluted and poured down the drain. Precipitate Ag^{+} as AgCl(*s*) by adding 1 drop of 12 M HCl to the $AgNO_3$ solution. After the precipitate has settled, decant the liquid into a beaker, dilute it with water, and flush the diluted solution down the drain. Transfer the AgCl slurry to the collection bottle labeled "waste AgCl." Dispose of the AgCl generated in the experiment in the same manner.

Decant the supernatant liquid from the half-cell containing AgI; dilute it and flush the diluted solution down the drain. Transfer the AgI slurry to the collection bottle labeled "waste AgI."

Precipitate Pb^{2+} as PbI_2. Decant and dilute the supernatant liquid and flush the diluted solution down the drain. Transfer the slurry of PbI_2 to the collection bottle marked "waste PbI_2." Dispose of the PbI_2 generated in the experiment in the same manner.

Table 22.1 Materials Needed for the Evaluation of Equilibrium Constants

K	*Electrode*	*Mixture*
K_{sp} for AgI	Ag	5.0 mL of 0.1 M $AgNO_3$ 1.0 mL of 1.0 M KI
K_{sp} for PbI_2	Pb	3.0 mL of 0.1 M $Pb(NO_3)_2$ 1.0 mL of 1.0 M KI
K_f for $Ag(NH_3)_2^{2+}$	Ag	3.0 mL of 0.1 M $AgNO_3$ 3.0 mL of 3.0 M NH_3
K_f for $Cu(NH_3)_4^{2+}$	Cu	3.0 mL of 0.1 M $Cu(NO_3)_2$ 3.0 mL of 3.0 M NH_3
K_f for $Zn(NH_3)_4^{2+}$	Zn	3.0 mL of 0.1 M $Zn(NO_3)_2$ 3.0 mL of 3.0 M NH_3

name ______ section ______ date ______

Pre-Lab Exercises for Experiment 22

These exercises are to be completed after you have read the experiment but before you come to the laboratory to perform it.

1. An electrochemical cell is constructed by coupling a Cu^{2+}/Cu half-cell with an Ag^{+}/Ag half-cell. The concentrations of Cu^{2+} and Ag^{+} are each 0.010 M. The observed potential, $\Delta\varepsilon$, for the cell is 452 mV, with Cu at the anode. If $\varepsilon° = 0.80$ V for the half-reaction $Ag^{+}(aq) + e^{-} \rightleftharpoons Ag(s)$, find the value of $\varepsilon°$ for the half-reaction: $Cu^{2+}(aq) + 2e^{-} \rightleftharpoons Cu(s)$.

2. When 5.0 mL of 0.10 M $AgNO_3$ and 0.50 mL of 12 M HCl are mixed, a precipitate of AgCl forms. Assume the reaction below goes to completion, and find the value of $[Cl^{-}]$ in the mixture.

$$Ag^{+}(aq) + Cl^{-}(aq) \rightarrow AgCl(s)$$

3. Silver ions are readily complexed by ammonia molecules according to the reaction:

$$Ag^+ + 2NH_3 \rightarrow Ag(NH_3)_2^+$$

a. Find the values of $[NH_3]$ and $[Ag(NH_3)_2^+]$ when 3.0 mL of 0.10 M $AgNO_3$ and 3.0 mL of 3.0 M NH_3 are mixed. Assume the complexation reaction goes to completion.

b. A cell is constructed in which the anode has a silver wire in contact with a mixture of 3.0 mL of 3.0 M NH_3 and 3.0 mL of 0.10 M $AgNO_3$, while the cathode has a silver wire in contact with 0.10 M $AgNO_3$. The potential of the cell is 505 mV. Find the value of K_f for $Ag(NH_3)_2^+$.

c. The K_f defined here is an "overall" formation constant. It is actually the product of two "step-wise" formation constants corresponding to the reactions:

$$Ag^+ + NH_3 \rightleftharpoons Ag(NH_3)^+$$

$$Ag(NH_3)^+ + NH_3 \rightleftharpoons Ag(NH_3)_2^+$$

If a significant portion of the silver ion is complexed as $Ag(NH_3)^+$, would your calculated value of K_f be too high or too low? Explain your answer briefly.

name ______ section ______ date ______

Summary Report on Experiment 22

A. *Determination of Standard Reduction Potentials*

Cell	*Electrode 1*	*Electrode 2*	*Meter Reading*	$\Delta\varepsilon$ (mV)	$\Delta\varepsilon$ (V)	$\Delta\varepsilon^\circ$ (V)
1	Ag	Zn (A)	______	______	______	______
2	Ag	Cu	______	______	______	______
3	Ag	Sn	______	______	______	______
4	Ag	Pb	______	______	______	______
5	Zn	Cu	______	______	______	______
6	Zn	Sn	______	______	______	______
7	Zn	Pb	______	______	______	______
8	Cu	Cu	______	______	______	______
9	Cu	Sn	______	______	______	______
10	Sn	Pb	______	______	______	______

Be sure to note which electrode is at the anode in each cell. Use Equation 22.6 to determine the value of $\Delta\varepsilon^\circ$ from the value of $\Delta\varepsilon$.

Calculation of Standard Reduction Potentials

Half-Cell Reduction Reaction	*Experimental Value of* ε°	*Literature Value of* ε°
$Ag^+ + e^- \rightarrow Ag$	0.80 V	0.80 V
$Cu^{2+} + 2e^- \rightarrow Cu$	______	______
$Pb^{2+} + 2e^- \rightarrow Pb$	______	______
$Sn^{2+} + 2e^- \rightarrow Sn$	______	______
$Zn^{2+} + 2e^- \rightarrow Zn$	______	______

B. *Determination of Equilibrium Constants*

Evaluation of K_{sp} for AgCl

Cell	*Half-cell 1*	*Half-cell 2*	*Meter Reading*	$\Delta\varepsilon$ (mV)	$\Delta\varepsilon$ (V)	$\Delta\varepsilon^{\circ}$ (V)
1	Ag (0.1 M)	Ag (*X* M)	________	________	________	________
2	Ag (*X* M)	Zn (0.1 M)	________	________	________	________

Be sure to note which electrode is at the anode in each cell. Use your data from Part A to determine the value of $\Delta\varepsilon^{\circ}$ for the cell reaction written *as it occurs in part B.* Recall that $\Delta\varepsilon^{\circ} = 0$ for any concentration cell.

Concentration of Cl^-

Cell	*Cell Reaction*	$\Delta\varepsilon$, V	$[Ag^+]_{6\text{th boat}} = [X]$	*Value of K*
1	$Ag^+ \ (0.1\ M) + Ag \rightarrow Ag^+ \ (X\ M) + Ag$	________	________	________
2	$Ag^+ \ (X\ M) + Zn \rightarrow Zn^{2+} \ (0.1\ M) + Ag$	________	________	________

Evaluation of Another *K*

Mixture assigned ________________

K to be evaluated ________________

Meter reading ________________

$\Delta\varepsilon$, mV ________________

$\Delta\varepsilon$, V ________________

Cell reaction ________________

$[M^{n+}]_{6th\ \text{boat}}$ ________________

Concentration of Cl^-, I^-, or NH_3 ________________

Value of *K* ________________

Reduction Potentials: Micro-Voltaic Cells

EXP
22P

Materials

CBL 2 interface
TI Graphing Calculator
DataMate program
Voltage probe
Sand paper
Forceps

0.10 M solns of Zn^{2+}, Cu^{2+}, Ag^{+}, Sn^{2+}, Pb^{2+}
Strips of Zn, Cu, Ag, Sn, Pb foils
1 M KNO_3
Strips of filter paper
12 M HCl
1.0 M KI
3.0 M NH_3

(Substitute) Procedure

A. Measurement of standard reduction potentials

1. Obtain and wear goggles.

2. Plug the voltage probe into Channel 1 of the CBL 2 interface. Use the link cable to connect the TI Graphing Calculator to the interface. Firmly press in the cable ends.

3. Turn on the calculator and start the DataMate program. Press CLEAR to reset the program.

4. Set up the calculator and interface for the voltage probe.

a. If the calculator displays VOLTAGE (V) in CH 1, proceed directly to Step 5. If it does not, continue with this step to set up your sensor manually.

b. Select SETUP from the main screen.

c. Press ENTER to select CH 1.

d. Select the voltage probe you are using from the SELECT SENSOR menu.

e. Select OK to return to the main screen.

5. Crossing three strips of double-sided tape, make an "asterisk—*", on a cardboard backing. Press a weighing boat firmly onto the middle of the vertical strip of tape. Then press five additional weighing boats onto the cross strips of tape, creating an open honeycomb pattern (Figure 22.4). Place approximately 5 mL of 0.10 M KNO_3 in the center boat. Place approximately 5 mL of 0.10 M $AgNO_3$, 0.10 M $Cu(NO_3)_2$, 0.10 M $Pb(NO_3)_2$, 0.10 M $Zn(NO_3)_2$, and 0.10 M $SnCl_2$, respectively, in each of the outer boats, labeling the boats by showing the symbol for each cation on the cardboard backing to which the weighing boats are attached. Be sure that the contents of the various boats are not allowed to mix.

6. Obtain strips of Ag, Cu, Pb, Zn, and Sn foils, which will serve as electrodes. Lift each metal electrode with forceps and sand it to remove any oxide coating. Place one end of the electrode in the solution of the corresponding ion (that is, Zn in Zn^{2+}, etc.) to complete each half-cell. (Do not handle the electrodes with your fingers; use the forceps at all times.) You may bend the electrodes, if necessary. You may also use a small piece of masking tape to secure the electrodes to the weighing boat edge or the cardboard on which the boats are mounted.

7. Dip one end of a strip of filter paper into the solution of zinc nitrate. Place the other end of the filter paper in the solution of potassium nitrate. (You may use a wooden applicator stick, if necessary, to submerge the second end of the paper in the potassium nitrate solution.) The filter paper "bridges" can function effectively as "salt bridges" (conduits for spectator ions) only if the entire length of the filter paper is moistened by the solutions in which the paper is immersed.

8. Use separate strips of filter paper to link the solutions of copper(II) nitrate, lead nitrate, silver nitrate, and tin(II) chloride to the central boat containing the solution of potassium nitrate. Two strips of filter paper, connecting any two outer boats to the center boat, constitute the salt bridge of your electrochemical cell.

9. Attach the alligator clip of one on the electrical leads to the zinc electrode.

10. Wait 5 seconds, then record the reading on the meter. Use a label (or a piece of masking tape) to mark the electrical lead to which the zinc metal electrode is attached. The lead with the tape will always be the anode in any cell that gives a positive reading on the meter.

11. Move the clips to new electrodes. Wait 5 seconds before recording the reading. For each cell, note the identity of the electrodes involved, and state which is the anode. Remember that the leads should always be attached to the electrodes in a manner that produces an on-scale reading. Continue until all of the boats have been coupled in pairs.

12. Use the Nernst equation to find the value of $\Sigma°$ for the Zn/Zn^{2+} half-cell, the Cu/Cu^{2+} half-cell, the Sn/Sn^{2+} half-cell, and the Pb/Pb^{2+} half-cell, assigning the Ag/Ag^{+} half-cell an $\Sigma°$ value of 0.80 V.

13. When you are finished, select QUIT and exit the DataMate program.

14. Use forceps to remove each of the pieces of metal from the filter paper. Rinse each piece of metal with tap water. Dry it and return it to the correct container. Remove the filter paper from the forceps, and discard it as directed.

B. Evaluation of Equilibrium Constants

15. Insert a sixth weighing boat in the open space of your honeycomb. Fill this cell with 5.0 M of 0.10 M $AgNO_3$ and 0.50 mL of 12 M HCl. Immerse a freshly sanded piece of silver foil into this mixture and connect the sixth boat to the central boat, as usual, with a strip of filter paper. Attach the alligator clips to the two silver foils and wait 15 seconds before recording the reading on your pH meter. Note which half-cell is the anode.

16. Move the alligator clip from the silver foil that is immersed in silver nitrate to the zinc electrode. Wait 15 seconds before recording the reading on the pH meter. Note which half-cell is the anode.

17. Remove the weighing boat that contains the silver chloride from the honeycomb.

18. Clean (see Disposal of Reagents) and dry the boat. Place it in the honeycomb once again and fill it, as your instructor directs, using a foil and mixture of solutions specified in Table 22.1. Use the potential of the resulting concentration cell to find the value of your assigned *K*.

Disposal of Reagents

19. The $Zn(NO_3)_2$, $Cu(NO_3)_2$, $SnCl_2$, and KNO_3 solutions may be diluted and poured down the drain.

20. Precipitate the Ag^+ as AgCl(*s*) by adding 1 drop of 12 M HCl to the $AgNO_3$ solution. After the precipitate has settled, decant the liquid into a beaker, dilute it with water, and flush the diluted solution down the drain. Transfer the AgCl slurry to the collection bottle labeled "waste AgCl." Dispose of the AgCl generated in the experiment in the same manner. Decant the supernatant liquid from the half-cell containing AgI; dilute it and flush the diluted solution down the drain. Transfer the AgI slurry to the collection bottle labeled "waste AgI."

21. Precipitate Pb^{2+} as PbI_2. Decant and dilute the supernatant liquid and flush the diluted solution down the drain. Transfer the slurry of PbI_2 to the collection bottle marked "waste PbI_2." Dispose of the PbI_2 generated in the experiment in the same manner.

EXP
23

Enthalpy of Hydration of Ammonium Chloride

Laboratory Time Required Three hours.

Special Equipment and Supplies

Burets
Buret clamp
Dewar flask calorimeter
Thermometer
Mortar and pestle
pH indicating paper
Balance
Standardized 1 M HCl, hydrochloric acid
Aqueous NH_3, ammonia
Methyl orange indicator
Solid ammonium chloride, NH_4Cl
Ice

Objective

In performing this experiment, students will use a Born-Haber cycle and data from a titration and from two calorimetry studies to evaluate the sum of the enthalpy changes of hydration for gaseous ammonia and gaseous hydrogen chloride.

Safety

Aqueous ammonia is **caustic.** Hydrochloric acid is **corrosive.** Avoid splashing these chemicals on your skin, in your eyes, or in your mouth.

The Dewar flask and thermometer are easily broken, resulting in a possible hazard of cuts from sharp edges.

The Dewar flask may implode violently, throwing glass fragments for some distance. Wear safety goggles at all times.

Broken thermometers may contaminate the laboratory with spilled mercury, a poison. Consult your instructor for clean-up procedures.

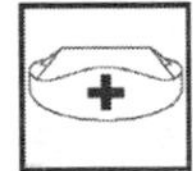

First Aid

If you have contact with an acid or base, flush the affected area thoroughly with water.

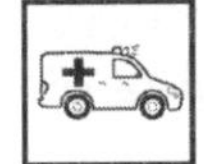

See a doctor if your eyes were exposed or if there is continued skin discomfort.

Minor cuts may be treated with antiseptic and covered with an adhesive bandage. Deep or extensive cuts require a doctor's attention.

Preamble

Calorimeters are used to study the heat absorbed or released by a system that undergoes a physical or a chemical change. A number of reliable procedures have been devised for quantitatively measuring the number of joules lost or gained by the reactants during this change. The ice calorimeter, for example, makes use of the fact that a known number of joules is needed to convert 1 g of ice at 0°C to water at the same temperature. The amount of ice melted, therefore, accurately determines the number of joules released during a reaction. Alternatively, the heat absorbed or released by the reaction system may be used to change the temperature of the calorimeter and its contents. If the temperature change and the heat capacity of the calorimeter and its contents are known, the quantity of heat involved may be calculated easily. In this experiment, the temperature change in a solution calorimeter will be used to study the enthalpy of a neutralization reaction and the enthalpy of solution of a salt.

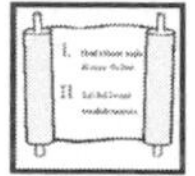

Principles

Chemical reactions are generally accompanied by a gain or loss of energy. This energy is very often in the form of heat. The amount of heat can be measured if the reaction is carried out in a calorimeter (a vessel so well insulated that its contents are an isolated "universe" consisting of a system (a chemical reaction) and its surroundings (usually the solution in which the reaction is taking place, the thermometer, a stirrer, and the inner walls of the vessel). Thus, the heat evolved by an exothermic reaction would equal the heat absorbed by the calorimeter and its contents. Similarly, the heat absorbed by an endothermic reaction would equal the heat lost by the calorimeter and its contents. These heat changes are represented by the symbol q. Obviously, q depends on the amount of reaction that takes place. If we evaluate q_{rxn} for 1 mole of reaction under constant pressure, q is then equal to ΔH, the enthalpy change for that reaction.

At times it is not convenient, or even possible, to measure the heat change associated with a reaction directly. In such cases, Hess' Law may be invoked. This principle states that, if a reaction can be regarded as the sum of two or more reactions, then the enthalpy change, ΔH, of the overall reaction is equal to the sum of the enthalpy changes of these contributing steps. The use of Hess' Law to evaluate the heat of formation of carbon monoxide is illustrated below. It is difficult to measure this reaction enthalpy directly because carbon monoxide tends to be oxidized further to carbon dioxide.

$$\begin{array}{ll} \text{C (graphite)} + O_2\ (g) \rightarrow CO_2\ (g) & \Delta H^\circ_{f,CO_2} = -394\ \text{kJ/mol} \\ CO_2\ (g) \rightarrow CO\ (g) + \frac{1}{2}O_2\ (g) & -\Delta H^\circ_{\text{combustion, CO}} = 284\ \text{kJ/mol} \\ \hline \text{C(graphite)} + \frac{1}{2}O_2\ (g) \rightarrow CO\ (g) & \Delta H^\circ_{f,CO} = -110\ \text{kJ/mol} \end{array}$$

A corollary of Hess' Law entails a cycle of thermochemical equations (a Born-Haber cycle). Because the net change in enthalpy in going around a cycle and returning to the starting point must be zero, the Born-Haber cycle is most often used when the enthalpies of all but one of the reactions in the cycle can be measured directly. The Born-Haber cycle to be investigated in this experiment is shown below.

$HCl(g)$	+	$NH_3(g)$	$\overset{\Delta H_1}{\rightarrow}$	$NH_4Cl(s)$
$\Delta H_5 \downarrow$		$\downarrow \Delta H_4$		$\downarrow \Delta H_2$
$HCl(aq, 1\ M)$	+	$NH_3(aq, 1M)$	$\overset{\Delta H_3}{\rightarrow}$	$NH_4Cl(aq, 0.5\ M)$

You will use this cycle to find the sum of the enthalpies of hydration of gaseous ammonia and gaseous hydrogen chloride ($\Delta H_4 + \Delta H_5$). You will determine two of the enthalpies experimentally. They are the enthalpy of solution of ammonium chloride (ΔH_2) and the enthalpy of neutralization of aqueous ammonia and aqueous hydrochloric acid (ΔH_3). The last enthalpy (ΔH_1) in the cycle is the enthalpy change for the synthesis of ammonium chloride crystals from gaseous ammonia and hydrogen chloride. You will not determine the value of ΔH_1 experimentally; instead, you will evaluate it from tabulated values of the enthalpies of formation of ammonium chloride crystals, hydrogen chloride gas, and ammonia gas. The standard **enthalpy of formation** of a substance is defined as the enthalpy change involved in the synthesis of a mole of the substance in its standard state from its constituent elements in their most stable forms under standard conditions (all pressures equal to 1 atm and all concentrations equal to 1 M). Values of enthalpies of formation are found in textbooks or tabulations of thermodynamic data.

You will use a Dewar flask as a calorimeter to evaluate the various enthalpies for the reactions in the cycle. You must first determine the heat capacity of the calorimeter (the number of joules — (J) — needed to raise the temperature of the calorimeter by 1°C). You will do so by mixing equal volumes of room-temperature water and cool water in the calorimeter and finding the final temperature of the mixture. In this case, the amount of heat lost by the room temperature water will equal the sum of the amount of heat gained by the cool water and the amount of heat gained by the calorimeter walls. This equality is illustrated in Equation 23.1.

$$-C_{H_2O}g_{H_2O}\left(t_f - t_r\right) = C_{H_2O}g_{H_2O}\left(t_f - t_c\right) + C_{cal}\left(t_f - t_c\right) \qquad (23.1)$$

where C_{H_2O} is the heat capacity of water in units of J/g-deg,
g_{H_2O} is the mass of cool water or room-temperature water,
t_r is room temperature,
t_c is the initial temperature of cool water,
t_f is the final temperature of the mixture, and
C_{cal} is the heat capacity of the calorimeter in units of J/deg.

Because the calorimeter does not come to thermal equilibrium immediately and because it is not perfectly insulated, the final temperature to be used in evaluating C_{cal} cannot be determined directly. Instead, you will mix the water samples and monitor the temperature change as a function of time. Plotting a graph of temperature versus time, and extrapolating back to the initial time of mixing, gives the value of t_f (Figure 23.1).

Once the heat capacity of the calorimeter is known, it must be taken into account in evaluating the heats of the reactions taking place in the vessel. The heat associated with the neutralization reaction between $HCl(aq)$ and $NH_3(aq)$ may be obtained from Equation 23.2, where C_{NH_3}, C_{HCl}, and C_{cal} are the heat capacities of the aqueous ammonia, hydrochloric acid, and calorimeter, respectively; g_{NH_3} and g_{HCl} are the masses of the solutions in containing the ammonia and hydrochloric acid; and Δt is the temperature change that results from the neutralization reaction. A typical plot for determining Δt is shown in Figure 23.2. Note that, if the neutralization process is exothermic, q_{neut} will be negative and the calorimeter will absorb heat, resulting in a temperature increase for the calorimeter and a positive value for q_{cal}.

$$q_{neut} = -q_{cal} = -\left(C_{NH_3} g_{NH_3} + C_{HCl} g_{HCl} + C_{cal} \Delta t\right) \tag{23.2}$$

Similarly, the amount of heat associated with the dissolution of ammonium chloride in water may be obtained from Equation 23.3, where C_{NH_4Cl} and g_{NH_4Cl} are the heat capacity and mass of the solution containing the ammonium chloride and Δt is the temperature change associated with the dissolution of ammonium chloride. Recall, again, that the thermometer is part of the surroundings while the dissolution "reaction" is the system. Thus, q_{sol} and q_{cal} will have opposite signs.

$$q_{sol} = -q\text{cal} = -\left(C_{NH_4Cl} g_{NH_4Cl} + C_{cal} \Delta t\right) \tag{23.3}$$

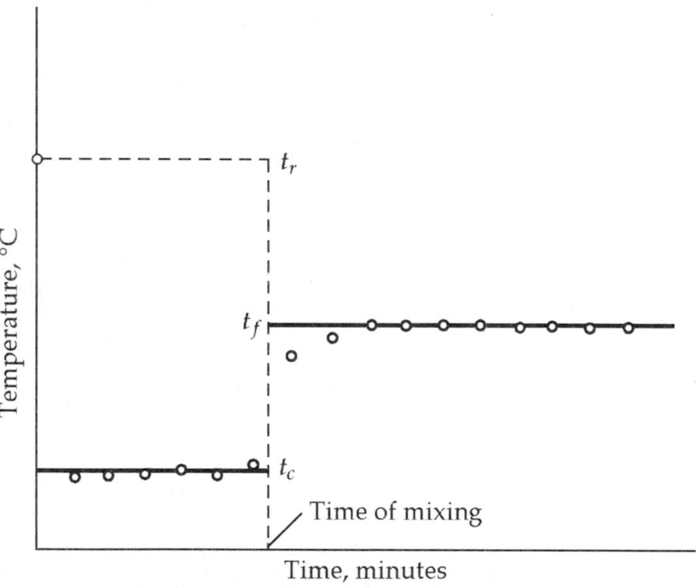

Figure 23.1 Determining the calorimeter constant

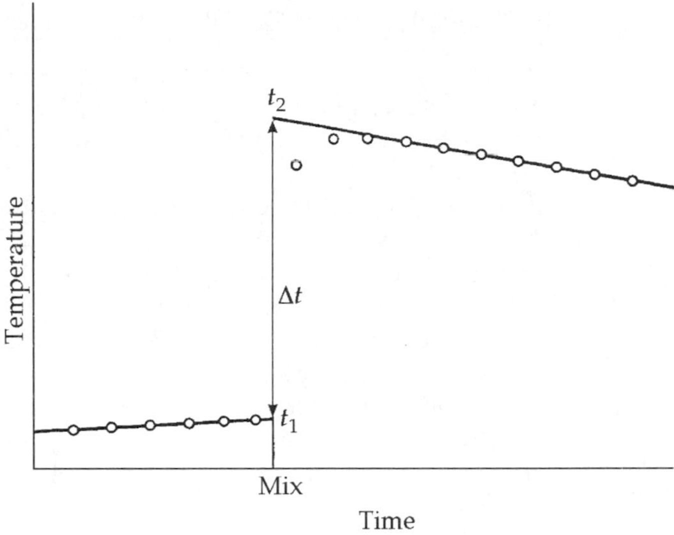

Figure 23.2 Typical time-temperature plot

Equations 23.2 and 23.3 may be greatly simplified if we make the following assumptions:

1. The heat capacities of all solutions of interest are the same (within 1 to 2%) as pure water at 25°C, 4.180 J/g·deg or 4.167 J/mL·deg, and are not temperature dependent.

2. The initial and final temperature (and therefore Δt) are the same for all solutions in a given reaction.

Making the appropriate changes in Equations 23.2 and 23.3 gives us:

$$q_{\text{neut}} = -\left(V_{\text{soln}} C_{\text{H}_2\text{O}} + C_{cal}\right)\Delta t \tag{23.4}$$

where V_{soln} is the sum of the volumes of the solutions containing the ammonia and the hydrochloric acid and $C_{\text{H}_2\text{O}} = 4.167 \text{ J/mL} \cdot \text{deg}$

$$q_{\text{sol}} = -\left(V_{\text{soln}} C_{\text{H}_2\text{O}} + C_{cal}\right)\Delta \text{t} \tag{23.5}$$

where V_{soln} is the volume of the solution containing the ammonium chloride and $C_{\text{H}_2\text{O}} = 4.167 \text{ J/mL} \cdot \text{deg}$.

The heats represented by q_{neut} and q_{sol} will have the units of joules. These heats are readily converted to enthalpies, in units of kJ/mole, by applying the appropriate conversion factor and considering the number of moles of reactants involved in the respective reactions.

Procedure

Procedure in a Nutshell

Determine the calorimeter constant. Standardize aqueous ammonia via titration with hydrochloric acid that was previously standardized for you. Measure the change in temperature resulting from the reaction of ammonia and hydrochloric acid in a calorimeter, and evaluate the enthalpy change of neutralization (ΔH_{neut}) for the reaction. Measure the change in temperature when ammonium chloride is dissolved in water in a calorimeter and evaluate the enthalpy change of solution (ΔH_{sol}) for NH_4Cl. Look up the enthalpy changes of formation (ΔH_f) for ammonia gas, hydrogen chloride gas, and solid ammonium chloride. Use all of the values thus obtained in a Born-Haber cycle to find the sum of the enthalpy changes of hydration for ammonia gas and hydrogen chloride gas.

Determining the Heat Capacity of the Calorimeter

Place roughly 250 mL of water in a 400-mL beaker and cool this water to 4°C or 5°C by placing the beaker in an ice bath. Use a clean, dry, 200-mL volumetric flask to transfer 200.00 mL of the cold water to the clean, dry calorimeter. Rinse the volumetric flask with room temperature water and then measure out exactly 200.00 mL of this water. Monitor the temperature of the water in the volumetric flask. Do not proceed until this temperature is constant and equal to room-temperature. Record this temperature. Assemble the calorimeter and stir the cool water, recording its temperature every 30 seconds, for 3 minutes. Add all of the room-temperature water to the cool water in the calorimeter as quickly as possible, while stirring. Record the exact time of mixing. Continue stirring. Record the temperature every 15 seconds for 5 minutes. Prepare a plot, similar to the one shown in Figure 23.1, and calculate the heat capacity of the calorimeter.

Standardization of Aqueous Ammonia

Record the exact concentration of the standardized HCl. Rinse and fill one buret with the HCl solution. Rinse and fill the other buret with the aqueous ammonia. Record the initial reading of each buret. Dispense 25-mL of HCl (read the buret to the nearest 0.01 mL both times) into a clean (but not necessarily dry) Erlenmeyer flask. Record the final buret reading. Add 25 mL of distilled water and 3 drops of methyl orange indicator to the acid in the flask. Titrate the acid solution with the aqueous ammonia until the indicator color changes from orange to yellow. Record the final reading for the buret containing the aqueous ammonia.

Determining the Heat of Neutralization of Ammonia and Hydrochloric Acid

Dry the interior of the calorimeter with a cloth towel or rag.

Caution!

THE DEWAR FLASK MAY IMPLODE VIOLENTLY IF BUMPED OR SCRATCHED.

Rinse the volumetric flask with several small portions of aqueous ammonia. Then fill the flask with NH_3 solution to the mark. Transfer the aqueous ammonia as completely as possible to the calorimeter. Record the temperature of the ammonia solution. Rinse the thermometer with distilled water to remove any base remaining on it. Rinse the volumetric flask with several small portions of distilled water. Then rinse it again with several small portions of hydrochloric acid. Fill the flask to the mark with HCl solution. Check the temperature of the acid in the volumetric flask. Rinse the exterior of the flask with hot or cold water, as necessary, to bring the temperature of the hydrochloric acid to within ±0.1°C of the temperature of the aqueous ammonia. Rinse the thermometer with distilled water to remove any acid remaining on it. Dry the thermometer. Monitor the temperature of the aqueous ammonia, recording its value every 30 seconds for 3 minutes. Add all of the hydrochloric acid to the aqueous ammonia in the calorimeter as quickly as possible, while stirring. Record the exact time of mixing. Continue stirring. Record the temperature every 15 seconds for 5 minutes. Prepare a plot, similar to the one shown in Figure 23.2, and calculate q_{neut}.

Determining the Heat of Solution of Ammonium Chloride

Using a mortar and pestle, powder 11.0 g of pure, dry NH_4Cl and transfer it to a small (clean, dry) beaker. Weigh the beaker and its contents. Clean and dry the calorimeter **carefully.** Use the volumetric flask twice to add 400.00 mL of distilled water to the calorimeter. Monitor the temperature of the water, recording its value every 30 seconds for 3 minutes. While stirring, add the ammonium chloride to the water in the calorimeter. Record the exact time of mixing. Continue stirring. Record the temperature every 15 seconds for 5 minutes. Weigh the empty beaker. Determine and record the mass of ammonium chloride used. Prepare a plot, similar to the one shown in Figure 23.2, and calculate q_{sol}.

Use your data to find the values of ΔH_2 (enthalpy of solution of ammonium chloride) and ΔH_3 (enthalpy of neutralization of ammonia and hydrochloric acid). Combine these with the value of ΔH_1 (enthalpy of synthesis of ammonium chloride) to find the sum of ΔH_4 and ΔH_5 (sum of the enthalpies of hydration of hydrogen chloride and ammonia). Note that ΔH_1 is to be calculated using standard enthalpies of formation. Cite your source for these data in your report. Append your plots, calculations, and results to your Summary Report.

Disposal of Reagents

All solutions may be neutralized (if necessary), diluted, and flushed down the drain.

name section date

Pre-Lab Exercises for Experiment 23

These exercises are to be completed after you have read the experiment but before you come to the laboratory to perform it.

1. Find the values for the enthalpy changes of formation of NaOH(*aq*), HCl(*aq*), KOH(*s*), and KOH(*aq*) in an appropriate reference work.

2. Write balanced chemical equations corresponding to the values found in answer to Question 1.

3. The enthalpy change for any reaction can be obtained by subtracting the sum of the enthalpy changes of formation of the reactants from the sum of the enthalpy changes of formation of the products. Use this method to obtain the value of ΔH_{neut} for the reaction of NaOH(*aq*) and HCl(*aq*) and the value of the enthalpy change of solution for KOH.

name ______________________ section ______________________ date __________

Summary Report on Experiment 23

Standardization of Aqueous Ammonia

Molarity of HCl	________
Final buret reading, HCl	________
Initial buret reading, HCl	________
Volume of HCl used	________
Final buret reading, NH_3	________
Initial buret reading, NH_3	________
Volume of NH_3 used	________

Determining the Heat Capacity of the Calorimeter

Room temperature		________
Temperature of cool water	*Time*	*Temperature*
	________	________
	________	________
	________	________
	________	________
	________	________
	________	________
Initial time of mixing		________

Time-temperature data:

Time	*Temperature*	*Time*	*Temperature*

Temperature of cold water __________

Temperature of warm water __________

Temperature of mixture __________

The calculated value of the calorimeter constant C_{cal} __________

name ______ section ______ date ______

Determining the Heat of Neutralization of Ammonia and Hydrochloric Acid

Temperature of hydrochloric acid ______

Temperature of aqueous ammonia

Time	*Temperature*
______	______
______	______
______	______
______	______
______	______
______	______

Initial time of mixing ______

Time-temperature data:

Time	*Temperature*	*Time*	*Temperature*
______	______	______	______
______	______	______	______
______	______	______	______
______	______	______	______
______	______	______	______
______	______	______	______
______	______	______	______
______	______	______	______
______	______	______	______
______	______	______	______

Determining the Heat of Solution of Ammonium Chloride

Mass of beaker plus NH_4Cl ______________

Mass of beaker ______________

Mass of NH_4Cl used ______________

Temperature of water

Time	Temperature

Initial time of mixing ______________

Time-temperature data:

Time	Temperature	Time	Temperature

EXP
24

Thermodynamic Prediction of Precipitation Reactions

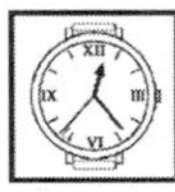

Laboratory Time Required

One hour. May be combined with Experiment 22.

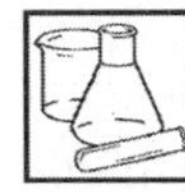

Special Equipment and Supplies

Dropper bottles

0.5 M solutions containing Ag^+, Ba^{2+}, Na^+, Ca^{2+}, Pb^{2+}, Cl^-, NO_3^- and SO_4^{2-} ions

Objective

In performing this experiment, the student will compare the predictions of thermodynamic calculations on precipitation reactions with observations made in the laboratory.

Safety

As usual, when chemicals are used in the laboratory, safety glasses should be worn and chemicals should not be ingested.

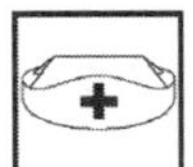

First Aid

Wash chemicals off your skin and out of your eyes with copious amounts of water.

Introduction

The chemistry curriculum has been criticized for being short on descriptive chemistry and long on theory — resulting in students who are often unfamiliar with basic phenomena and uncomfortable with chemical principles. This experiment attempts to overcome both problems by requiring you to work directly with thermodynamic concepts and calculations and to confirm your calculated predictions by direct observation of several precipitation reactions.

Principles

A spontaneous change is one that occurs by itself, without the exertion of any outside force. A mixture of hydrogen and oxygen gas changes spontaneously (and explosively) into water after being ignited by a spark. Iron rusts spontaneously, albeit slowly, when it is exposed to air and water.

Many spontaneous chemical changes, such as combustion of hydrocarbons, are exothermic. However, there are many examples of endothermic processes that occur spontaneously. These include the melting of ice at ambient temperatures above 0°C and the boiling of water at 100°C (at 760 torr of pressure). In both of these cases, spontaneous change occurs in the direction of a less-ordered state.

Two state functions have been defined to describe the tendency for a change to occur spontaneously. The first of these is ΔH, the enthalpy change. A negative value for ΔH denotes an exothermic process and is a factor that favors spontaneous change. The second state function is ΔS, the entropy change. Entropy is a measure of randomness. A positive value for ΔS denotes that the change will result in a more random (less ordered) system, a factor that also favors spontaneous change.

Very often the direction of spontaneous change is determined by temperature. For instance, water changes spontaneously into ice when placed in a freezer ($t < 0°C$), but ice cubes melt spontaneously when removed from a freezer ($t > 0°C$). In these cases, the signs of ΔH and ΔS work in opposition (e.g., ΔH is favorable when water freezes, but ΔS is unfavorable; ΔH is unfavorable when ice melts, but ΔS is favorable), and temperature is the factor that determines whether ΔH or ΔS will dominate. This information is incorporated into a single state function, the Gibbs free energy, ΔG, defined (for constant temperature systems) in Equation 24. 1.

$$\Delta G = \Delta H - T\Delta S \tag{24.1}$$

It is easy to see that ΔG must be negative if a process is spontaneous because a change for which ΔH is negative and ΔS is positive will surely be spontaneous. Endothermic processes that give positive entropy changes will be spontaneous at high temperatures (where $T\Delta S$ dominates). Exothermic processes that give negative entropy changes will be spontaneous at low temperatures (where ΔH dominates).

Many textbooks give tabulations of values for specific free-energy changes ($\Delta G^\circ_{f,298}$), enthalpy changes ($\Delta H^\circ_{f,298}$), and entropies ($S^\circ_{f,298}$). The ° symbols indicate that the values are being given for changes involving substances in their standard states (all partial pressures are 1 atm and all molarities are 1 M). The *f* subscript on ΔG and ΔH denotes formation; ΔG_f° and ΔH_f° are the free energy and enthalpy changes, respectively, which are associated with a reaction in which one mole of product is formed from its constituent elements in their most stable forms under standard conditions. The 298 subscript indicates that all quantities have been corrected to the values they would have if the change were to occur at 298 K.

Table 24.1 shows how to use tabulated values of $\Delta G^\circ_{f,298}$, $\Delta H^\circ_{f,298}$, and S°_{298} for carbon, oxygen, and carbon dioxide to calculate the free energy change in the combustion of carbon.

Table 24.1 Calculating the Free Energy Change for the Combustion of Carbon (graphite)

	C (graphite, 298)	+	O_2 (*g*, 298)	→	CO_2 (*g*, 298)
$\Delta H^\circ_{f,298}$	0		0		–393.5 kJ/mol
S°_{298}	5.73 J/mol·K		205.0 J/mol·K		213.6 J/mol·K
$\Delta G^\circ_{f,298}$	0		0		–394.3 kJ/mol

$$\Delta H^\circ_{rxn,298} = \Delta H^\circ_{f,298}(CO_2) - \Delta H^\circ_{f,298}(O_2) - \Delta H^\circ_{f,298}(C)$$

$$\Delta H^\circ_{rxn,298} = -393.5 \text{ kJ/mol}$$

$$\Delta S^\circ_{rxn,298} = S^\circ_{298}(CO_2) - S^\circ_{298}(O_2) - S^\circ_{298}(C)$$

$$\Delta S^\circ_{rxn,298} = 2.9 \text{ J/mol}\cdot\text{K}$$

$$\Delta G^\circ_{rxn,298} = \Delta H^\circ_{rxn,298} - (298)\Delta S^\circ_{rxn,298} = -394.4 \text{ kJ/mol}$$

$$\Delta G^\circ_{rxn,298} = \Delta G^\circ_{f,298}(CO_2) - \Delta G^\circ_{f,298}(O_2) - \Delta G^\circ_{f,298}(C)$$

$$\Delta G^\circ_{rxn,298} = -394.3 \text{ kJ/mol}$$

Note that you can obtain the standard free energy change for any reaction, $\Delta G^\circ_{rxn,298}$, by subtracting the sum of the $\Delta G^\circ_{f,298}$'s for the reactants from the sum of $\Delta G^\circ_{f,298}$'s for the products (see Equation 24.2).

$$\Delta G^\circ_{rxn,298} = \sum_{\text{products}} G^\circ_{f,298} - \sum_{\text{reactants}} G^\circ_{f,298} \tag{24.2}$$

The combustion of carbon to give carbon dioxide is, of course, a spontaneous process. This is confirmed by the fact that ΔG°_{298} for the combustion process is –394.4 kJ/mol. This means that when one mole of graphite, the most stable form of carbon, is combined with one mole of oxygen, at one atmosphere pressure, to produce one mole of CO_2, at one atmosphere pressure, the free energy of the system decreases by 394.4 kJ. If the reaction is not performed with all materials in their standard states, if the reaction temperature is not 298 K, if the pressure of the either of the gases is not one atmosphere, or if more or less than one mole of carbon is consumed in the reaction, then a value of ΔG_{rxn} will be obtained that will differ from ΔG°_{rxn}. The relationship between ΔG_{rxn} and ΔG°_{rxn} is shown in Equation 24.3.

$$\Delta G_{rxn} = \Delta G_{rxn}{}^\circ + RT \ln Q \tag{24.3}$$

In Equation 24.3, R is the ideal gas constant (8.314 J/K·mol), T is the Kelvin temperature, and Q denotes the reaction quotient. As is customary for evaluating equilibrium constants and reaction quotients, liquids, solids, and solvents are represented by unity; solute concentrations are represented by molarity; and the pressures of gases are given in atmospheres.

In this experiment, you will be calculating the ΔG_{rxn} for a variety of possible precipitation reactions. You will use your calculated values to predict whether a precipitate will form when two solutions are mixed.

Then you will actually mix the reagents and attempt to confirm your predictions. You will also calculate ΔG_{rxn} for the precipitation of a few salts at different temperatures. You will place the precipitates in hot or cold baths and attempt to confirm your predictions regarding the change in solubility as a function of changing temperature as well. Tables 24.2 and 24.4 list the data you will use to make your calculations. The following paragraphs explain how to use this data.

Table 24.2 Gibbs Free Energies ($\Delta G^{\circ}_{f,298}$) for Ions in 1 M Solution and Solids

Anions → ↓ Cations	Cl^- –131.228	I^- –51.57	NO_3^- –108.74	SO_4^{2-} –744.53
Ag^+ 77.107	–109.789	–66.19	–33.41	–618.41
Ba^{2+} –560.77	–1296.32 W2	–	–796.59	–1362.2
Na^+ –261.905	–384.138	–286.06	–367.00	–3646.85 W10
Ca^{2+} –553.58	–748.1	–528.9	–743.07	–1797.28 W2

Table 24.3 Calculating ΔG for the Precipitation of AgCl

Ag^+ (0.25 M) + Cl^- (0.25 M) → AgCl (*s*)

$$\Delta G^{\circ}_{rxn,298} = \Delta G^{\circ}_{f,298}(\mathrm{AgCl}) - \Delta G^{\circ}_{f,298}(\mathrm{Ag^+}) - \Delta G^{\circ}_{f,298}(\mathrm{Cl^-})$$

$$\Delta G^{\circ}_{rxn,298} = -109.789 - (77.107) - (-131.228)\ \mathrm{kJ/mol}$$

$$\Delta G^{\circ}_{rxn,298} = -55.668\ \mathrm{kJ/mol}$$

$$\Delta G_{rxn,298} = \Delta G^{\circ}_{rxn,298} + 2.303RT\log Q$$

$$\Delta G_{rxn,298} = -55.668\ \mathrm{kJ/mol} + 2.303\frac{(8.314\ \mathrm{J/mol\cdot K})(298\ \mathrm{K})}{1000\ \mathrm{J/kJ}}\log\frac{1}{(0.25)(0.25)}$$

$$\Delta G_{rxn,298} = -48.797\ \mathrm{kJ/mol}$$

The row and column headings of Table 24.2 show the standard free energies of formation of the various ions under consideration, with the standard state as a 1 M solution. Thus, $\Delta G^{\circ}_{f,298}$ for Ag^+ is 77.107 kJ/mol and $\Delta G^{\circ}_{f,298}$ for Cl^- is –131.228 kJ/mol. Entries within the body of the table show the standard free energies of formation for the crystalline solids that result from the combinations of the various ions whose rows and columns intersect to create the compound's cell. For instance, $\Delta G^{\circ}_{f,298}$ for AgCl is –109.789 kJ/mol. The "W10" entry in the cell corresponding to sodium sulfate indicates that the most likely precipitate is $Na_2SO_4 \cdot 10H_2O$. When you write the equation for the precipitation of such a hydrated salt, water will appear as a reactant. Therefore, in the calculation of ΔG°_{rxn} you will need to consider $\Delta G^{\circ}_{f,298}$ for water, which has a value of –237.129 kJ/mol.

Because you will not be working with 1 M solutions, the free-energy changes you will be calculating will not be standard free energies; rather, Equation 24.3 will be needed to convert the ΔG^{o}'s to ΔG's. Table 24.3 shows the calculation of ΔG for the possible reaction between Ag^{+} ions and Cl^{-} ions to form the precipitate, silver chloride. Because $\Delta G_{rxn,298}$ for the precipitation of AgCl is negative, it is predicted that the precipitate will form when equal volumes of 0.5 m solutions of Ag^{+} and Cl^{-} ions are mixed.

The ΔG^{o}'s and ΔG's you have calculated so far were evaluated at 298 K. There will be times when you may wish to evaluate these functions at other temperatures. This is easily accomplished because the values of ΔG^{o} and ΔS^{o} are relatively independent of temperature. Thus, ΔG^{o} can be evaluated at any temperature by the use of Equation 24.4, where T is the Kelvin temperature and ΔH^{o}_{298} and ΔS^{o}_{298} are the enthalpy and entropy change, respectively, for the reaction.

$$\Delta G^{o}_{T} = \Delta H^{o}_{298} - T\Delta S^{o}_{298} \tag{24.4}$$

Values of ΔH^{o}_{298} and S^{o}_{298} for several ions and crystalline solids are given in Table 24.4. Use the values given to decide whether precipitates would form when equal volumes of 0.5 M solutions of Pb^{2+} and Cl^{-} are mixed at 273 K, 298 K, and 373 K. Do similar calculations for mixtures of equal volumes of 0.5 M solutions Ba^{2+} and NO_3^{-}. Use Equation 24.4 to evaluate ΔG^{o}_{T} at $T = 273$ K, 298 K, and 373 K. Use Equation 24.3 to evaluate ΔG_{T} at those temperatures. An example of this kind of calculation is given in Table 24.5. Because ΔG_{rxn} is negative for the precipitation reaction, it is predicted that a precipitate will form if equal volumes of 0.5 M Pb^{2+} and 0.5 M I^{-} are mixed and cooled to 273 K.

Table 24.4 Values of $\Delta H^{o}_{f,298}$ and S^{o}_{298} for Various Ions and Solids

	$\Delta H^{o}_{f,298}$, *kJ/mol*	S^{o}_{298}, *J/K·mol*
Pb^{2+}	–2	10
Cl^{-}	–167	56
$PbCl_2$	–359	136
I^{-}	–55	111
PbI_2	–175	175
Ba^{2+}	–538	10
NO_3^{-}	–205	146
$Ba(NO_3)_2$	–992	214

Table 24.5 Calculating ΔG for the Precipitation of PbI_2 at 273 K

$$Pb^{2+}\ (0.25\ M) + 2I^-\ (0.25\ M) \rightarrow PbI_2\ (s)$$

$$\Delta H^\circ_{rxn,298} = -175 - (-2 + 2(-55)) = -63\ kJ$$

$$\Delta S^\circ_{rxn,298} = 175 - (10 + 2(111)) = -57\ J/K$$

$$\Delta G^\circ_{rxn,273} = -63000 - 273(-57) = -47 \times 10^3$$

$$\Delta G_{rxn,273} = -47 \times 10^3\ J + 2.303\left(8.314\ \frac{J}{K}\right)(273\ K)\log\frac{1}{(0.25)(0.25)^2}$$

$$\Delta G_{rxn,273} = -47 \times 10^3\ J + 6.3 \times 10^3\ J = -41 \times 10^3\ J$$

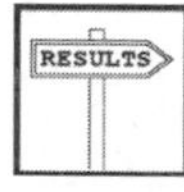

Procedure

Procedure in a Nutshell

Calculate the value of $\Delta G_{rxn,298}$ for each possible precipitate that could result from the mixing of solutions of the ions listed in Table 24.2. Then, mix the solutions and see how accurate your predictions were. Calculate the value of $\Delta G_{rxn,T}$ for the precipitation of both barium nitrate and lead chloride at low temperature and at high temperature. Again, do the mixing and see if your observations agree with the predictions of the calculations.

Calculate the value of $\Delta G_{rxn,298}$ for each solid that might result when each of the 0.5 M solutions of cations listed in Table 24.2 is mixed with an equal volume of each of the anion solutions listed. Note that mixing equal volumes of 0.5 M solutions of cations and anions will result in 0.25 M solutions after mixing. Record your values of the ΔG's in the upper space of the boxes on the Summary Report Sheet. Then mix the solutions and note in the lower spaces whether your observations confirm (C) or deny (D) your predictions. Below the table of results, note the appearance of the precipitates you observe and briefly discuss possible reasons for any discrepancies between your predictions and your observations.

Also, calculate the values of $\Delta G_{rxn,T}$ for the precipitation of $PbCl_2$ and $Ba(NO_3)_2$ from solutions that are 0.25 M in lead ions and chloride ions and 0.25 M in barium ions and nitrate ions, respectively, for T = 273 K, 298 K, and 373 K. Mix the appropriate solutions at room temperature. If no precipitate results, cool the mixture in an ice bath. If a precipitate does appear, heat the test tube containing the precipitate and supernatant liquid in a boiling water bath.

Note whether the precipitate dissolves and whether your observations are in accord with your calculations. If they are not, briefly discuss possible reasons for the discrepancies.

Disposal of Reagents

The small quantities of chemicals used in this experiment may be flushed down the drain with copious amounts of water.

name section date

Pre-Lab Exercises for Experiment 24

These exercises are to be completed after you have read the experiment but before you come to the laboratory to perform it.

1. Write the chemical equation corresponding to $\Delta G^\circ_{f,298}$ for $Na_2SO_4 \cdot 10H_2O$. How does this equation differ from the net ionic equation that shows $Na_2SO_4 \cdot 10H_2O$ precipitating when aqueous solutions of Na^+ and SO_4^{2-} are mixed?

2. The value of $\Delta G^\circ_{f,298}$ for Mg^{2+} is –454.8 kJ/mol. The value of $\Delta G^\circ_{f,298}$ for $MgSO_4 \cdot 7H_2O$ is –3383.76 kJ/mol. Will a precipitate form when equal volumes of 0.5 M Mg^{2+} and 0.5 M SO_4^{2-} are mixed?

name _______________ section _______________ date _______________

Summary Report on Experiment 24

	Cl^-	I	NO_3^-	SO_4^{2-}
Ag^+				
Ba^{2+}				
Na^+				
Ca^{2+}				

	$\Delta G_{rxn,273}$	$\Delta G_{rxn,298}$	$\Delta G_{rxn,373}$
$PbCl_2$	______	______	______
$Ba(NO_3)_2$	______	______	______

Observations when 0.5 M Pb^{2+} is mixed with 0.5 M Cl^-

Observations when 0.5 M Ba^{2+} is mixed with 0.5 M NO_3^-

EXP
25

Kinetic Study of the Reaction Between Iron(III) Ions and Iodide Ions

Laboratory Time Required

Three hours for well-prepared students, working in teams.

Special Equipment and Supplies

Constant-temperature baths
Burets
Timer
Thermometer, 0.1 subdivisions
Pipets
Pipet bulbs

0.04 M $Fe(NO_3)_3$, iron(III) nitrate in 0.15 M HNO_3, nitric acid
0.15 M HNO_3, nitric acid
0.04 M KI, potassium iodide
$Na_2S_2O_3 \cdot 5H_2O$ (*s*), sodium thiosulfate pentahydrate
0.2% starch solution
$Fe(NH_4)_2(SO_4)_2 \cdot 6H_2O$, iron(II) ammonium sulfate in solid form or as a 0.002 M solution in 0.15 M HNO_3
0.02 M standard KIO_3, potassium iodate
0.5 M H_2SO_4, sulfuric acid
Dilute $NaOH$, sodium hydroxide

Objective

In performing this experiment, students will use teamwork in obtaining the rate law for the reaction between iron(III) ions and iodide ions.

Safety

None of these solutions is very hazardous, but you should still exercise care to avoid chemical contact with your skin, mouth, or eyes.

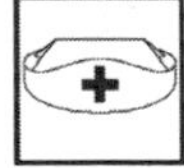

First Aid

Thoroughly flush the affected area with water.

Preamble

Much of chemistry is concerned with questions such as, "How much of one reactant is needed to consume another reactant?" or "How much product can be obtained from the reaction of the two reagents?" Such questions can be answered with simple stoichiometric calculations if the reaction goes smoothly to completion. However, many reactions are more complicated. It is possible that the product will have a tendency to decompose, regenerating the initial reactants. In that case, we can use thermodynamics to determine the extent of reaction. However, thermodynamics tells us only the direction of spontaneous change, not the rate at which the reaction will proceed. Kinetics is the branch of chemistry that concerns itself with rates of reaction. The aim of a kinetics experiment is the discovery of the rate expression (a relation between the rate of the reaction and such experimental variables as reactant concentrations, temperature, and the presence of catalysts). Knowledge of the rate law is useful if one desires to alter the rate (to speed up the production of a desired product or to slow down its decay) or wishes to probe the molecular basis for the reaction by considering possible mechanisms (sets of individual steps explaining in detail how the reactants are transformed into products).

Principles

Writing the rate expression for a reaction requires an experimental determination of the dependence of the rate of the reaction on the concentrations of the various reactants. In the Method of Initial Rates, a series of reaction mixtures is prepared such that any two mixtures differ only in the initial concentration of one reactant. Because the rate will probably change as the reactant under study is being used up, it is necessary to make the determination of the rate early in the reaction, before the concentrations of the reactants have changed significantly from their initial values.

The Rate Law

Consider the hypothetical reaction represented by Equation 25.1:

$$2A+B \rightarrow A_2B \tag{25.1}$$

The rate may be related to either the change in product concentration per unit time or the change in reactant concentration per unit time, according to Equation 25.2.

$$\text{rate} = -\frac{1}{2}\frac{d[A]}{dt} = -\frac{d[B]}{dt} = \frac{d[A_2B]}{dt} \tag{25.2}$$

Note that Equation 25.2 is not a rate expression. It is simply a "normalization"relation so that a value reported for the rate is not dependent on the manner in which the progress of the reaction was studied. For instance, when 1 mol L^{-1} of A is being consumed per minute, the value of [B] is decreasing by 0.5 mol L^{-1} min^{-1} and the value of $[A_2B]$ is increasing by 0.5 mol L^{-1} as well. Use of Equation 25.2 would give a reported rate of 0.5 mol L^{-1} min^{-1} whether the reaction was being monitored by observing the rate of consumption of either reactant or the rate of production of A_2B.

The rate expression (sometimes called a rate law) has the general form shown in Equation 25.3.

$$\text{rate} = k[A]^a[B]^b \tag{25.3}$$

where k is the specific rate constant for the reaction, and a and b represent the order of the reaction with respect to A and B. It should be emphasized that a and b are determined experimentally and are not deduced from the stoichiometry of the reaction. The order with respect to each reactant may be positive or negative, integral or fractional, or even zero.

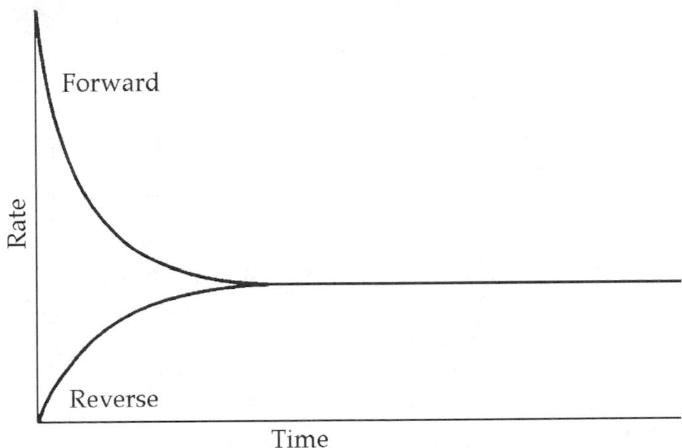

Figure 25.1 Change with time of the forward and reverse reaction rates for a reversible reaction

Reversible Reactions

The top curve of Figure 25.1 shows how a system initially containing only A and B behaves. The rate of the forward reaction is large at the start but decreases, as predicted from Equation 25.3, as the reactants are converted to A_2B. If the reaction is reversible, as assumed in Figure 25.1, the rate of the reverse reaction is related to the concentration of A_2B. The rate of the reverse reaction is zero at the start, when no A_2B is present, but increases with time as A_2B is formed by the forward reaction. Eventually, of course, the forward and reverse rates become equal, and the system is then at equilibrium.

We assumed in Equation 25.3 that the rate of formation of product is related to the forward reaction rate, according to the expression shown in Equation 25.4.

$$\frac{d[A_2B]}{dt} = \text{rate}_{(\text{forward})} \tag{25.4}$$

However, if the reaction is reversible, the expression shown in Equation 25.5 applies. The addition of the $\text{rate}_{(\text{reverse})}$ term greatly complicates the calculations, and we therefore try to work under conditions where the reaction is not reversible or where the reverse rate can be set equal to zero. An obvious advantage of the initial rate method is that the reverse rate is negligible under these conditions and, therefore, the reverse reaction can be ignored.

$$\frac{d[A_2B]}{dt} = \text{rate}_{(\text{forward})} - \text{rate}_{(\text{reverse})} \tag{25.5}$$

The Iron(III) Ion–Iodide Ion System

In this experiment, you will study the oxidation of I^- by Fe^{3+} ions, as shown in Equation 25.6. Note that the triiodide ion, I_3^-, is simply an iodine molecule complexed with an iodide ion. This complexation greatly increases the solubility of the iodine species in water.

$$2Fe^{3+} + 3I^- \rightarrow 2Fe^{2+} + I_3^- \tag{25.6}$$

The expected rate expression for the reaction of Fe^{3+} and I^- is given in Equation 25.7.

$$\text{rate} = -\frac{1}{2}\frac{d[Fe^{3+}]}{dt} = \frac{d[I_3^-]}{dt} = k[Fe^{3+}]^a[I^-]^b \tag{25.7}$$

Parts A and B of the experiment are concerned with evaluating the exponents *a* and *b*, respectively, in the rate expression. Part C is concerned with the effects of temperature on the reaction rate.

You will determine the initial rate by measuring the time, in seconds, required for part (about 4×10^{-5} mole) of the Fe^{3+} to be reduced to Fe^{2+}. Accomplishing this objective will require the mixing of a number of solutions, as will be explained below.

The Components of the Mixtures

The components of the mixtures necessary for the experiment are listed in Table 25.1. The mixtures are divided into two categories — one containing the iron(III) ions and the other containing the iodide ions. Each mixture also contains other substances, such as nitric acid, thiosulfate ions, starch, and water. The role of each component is discussed below.

Iron(III) ions are supplied by dissolving iron(III) nitrate in water. Their role is obvious; iron(III) is involved in the reaction to be studied.

Nitric acid is needed to suppress the hydrolysis, or reaction with water, of the iron(III) ions. (See Equation 25.8.) LeChatelier's principle tells us that iron(III) ions will not react with water to produce hydrogen ions in an acidic environment. We need to suppress this reaction so that the Fe^{3+} ions react only with the iodide ions in the reaction we are studying.

$$Fe^{3+} + H_2O \rightarrow FeOH^+ + H^+ \tag{25.8}$$

Water is used to bring the total volume of each of the mixtures to 50.00 mL. Having a constant volume for the mixtures allows us to vary the concentration of a single component by changing the volume of stock solution used in preparing the mixtures.

Iodide ions are prepared by dissolving potassium iodide in water. Like the Fe^{3+} ions, the I^- ions have an obvious role; they, too, are reactants in the system under study.

Thiosulfate ($S_2O_3^{2-}$) ions are supplied by dissolving sodium thiosulfate in water. Note that the stock solution of thiosulfate is ten times less concentrated than the stock solutions containing the iron(III) and iodide ions. This is a deliberate move that will permit us to evaluate the rate of the reaction before the initial concentrations of the reactants have changed significantly. To see how this works, we need to consider the next component of the mixture, starch.

Starch-containing solutions acquire a blue color in the presence of iodine (or triiodide ions). The appearance of a blue color will serve as an indication that the reaction has reached the point at which we want to evaluate the rate. We want this point to come early in the reaction but not instantaneously. Thus, we need both the thiosulfate ions and the starch to be present in the reaction mixture.

Triiodide ions, produced by reaction with iron(III), will not affect the color of the starch as long as the thiosulfate ions are present. This is so because thiosulfate ions react with triiodide ions to regenerate iodide ions. (See Equation 25.9.)

$$I_3^- + 2S_2O_3^{2-} \rightarrow 3I^- + S_4O_6^{2-} \qquad (25.9)$$

Only when the small amount of thiosulfate that is present in each mixture is consumed do the triiodide ions begin to accumulate and affect the color of the starch. The time that has elapsed between the mixing of Fe^{3+}-containing solution and the I^--containing solution is noted and the rate of the reaction is then evaluated.

As was the case with the mixtures that supplied the iron(III) ions, the mixtures that supply the iodide ions will also be kept at a constant volume through the addition of water. Again, this is done so that altering the volume of one stock solution changes the concentration of only one reactant under study. To illustrate why it is necessary to bring the overall mixtures to constant volume, consider what happens when 10.00 mL of 0.04 M Fe^{3+} and 10.00 mL of 0.04 M I^- are mixed; each solution dilutes the other so that the overall mixture is 0.02 M in Fe^{3+} and 0.02 M in I^-. If we wanted to study the effect of doubling the value of $[Fe^{3+}]$ and we simply mixed 20.00 mL of 0.04 M Fe^{3+} with 10.00 mL of 0.04 M I^-, the mutual dilution would produce a solution that is 0.027 M in iron(III) and 0.013 M in iodide ions. (We would have altered the concentrations of both ions from the values they had in the first mixture.) However, if we mix 10.00 mL of 0.04 M Fe^{3+} with 10.00 mL of 0.04 M I^- and 20.00 mL of water, we will find that mixture 1 contains both 0.01 M Fe^{3+} and 0.01 M I^-. Likewise, mixing 20.00 mL of 0.04 M Fe^{3+}, 10.00 mL of 0.04 M I^-, and 10.00 mL of water will produce a mixture that is 0.02 M Fe^{3+} but still 0.01 M I^-, as desired.

Defining the Rate

The development of the blue color in a reaction mixture occurs when all of the thiosulfate ions in the mixture have been consumed. Thus, the rate of the reaction can be defined as shown in Equation 25.10, where $[S_2O_3^{2-}]_0$ is the initial concentration of thiosulfate ions in the mixture and it is assumed that the time is zero seconds when the reactants are mixed.

$$\text{Experimental rate} = \frac{-\Delta[S_2O_3^{2-}]}{\Delta t} = \frac{-(0-[S_2O_3^{2-}])_i}{t-0} = \frac{[S_2O_3^{2-}]_i}{\text{time elapsed}} \qquad (25.10)$$

Although Equation 25.10 does provide a valid rate definition, it would be difficult to compare its results with rates defined in the more traditional manner (as a function of the rate of disappearance of one of the reactants or the rate of appearance of one of the products). Fortunately, it is easy to relate the rate of consumption of thiosulfate ions with the rate of consumption of iron(III) ions. Examination of Equations 25.6 and 25.9 reveals that two iron(III) ions are consumed for each triiodide ion produced and that each triodide ion, in turn, reacts with two thiosulfate ions. Therefore, we can define a normalized rate for our reaction (see Equation 25.11) that can be compared to other studies of the Fe^{3+}/I^{-} system.

$$\Delta[Fe^{3+}] = \Delta[S_2O_3^{2-}];\ \text{rate} = \frac{-1/2d\ [S_2O_3^{2-}]}{dt} = \frac{-1/2d\ [Fe^{3+}]}{dt} \qquad (25.11)$$

In Part C, the effects of temperature variations will be studied. The effect of temperature on reaction rate is treated in your text theoretically. Therefore, we will deal with the subject only briefly here. The specific rate constant, k, is related to the Kelvin temperature T by the expression shown in Equation 25.12, where E_a is the activation energy for the reaction. The activation energy represents the minimum energy required for the reactants to pass over an energy barrier to form products, as shown in Figure 25.2. Because the reactant particles have a statistical distribution of energies, at a given temperature only a fraction of the particles will have sufficient energy to react upon colliding. If the absolute temperature is increased, however, the fraction of the particles having the energy needed to react upon colliding also increases, resulting in a greater rate or a larger value for the specific rate constant.

$$k = Ae^{-E_a/RT} \qquad (25.12)$$

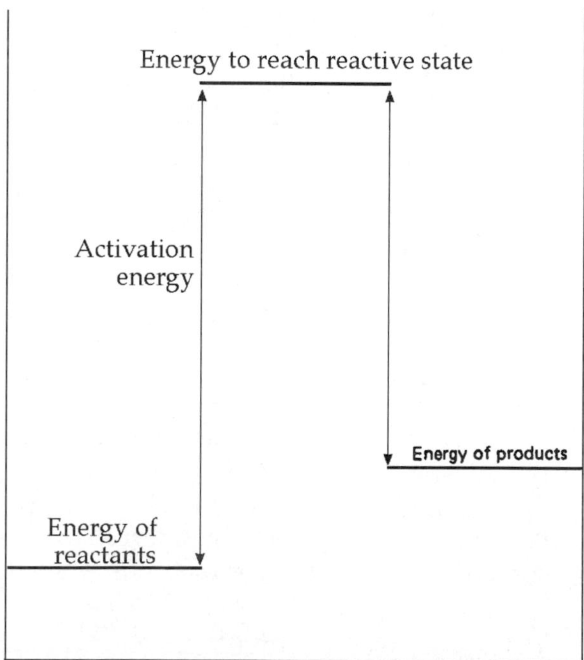

Figure 25.2 Schematic representation of the energy barrier for a chemical reaction

If A is a constant in Equation 25.12, then expressions for k as a function of T may be derived. (See Equations 25.13 and 25.14.)

$$\ln k = -\frac{E_a}{R} + \text{constant} \tag{25.13}$$

$$\Delta \ln k = -\frac{E_a}{R}\Delta(\frac{1}{T}) \tag{25.13}$$

A plot of $\ln k$ (or $2.303\log\, k$) versus $1/T$ should therefore be linear, with a slope equal to $-E_a/R$.

A Word on Logistics

Many solutions must be used in preparing each reaction mixture to be studied. The thiosulfate solution is somewhat unstable and is best prepared on the day it is to be used. The rate of reaction is quite sensitive to changes of temperature. All of these factors can affect the accuracy of the results obtained in this experiment. Very satisfactory results are obtained efficiently if the class is divided into teams, each with a specific task to perform that will facilitate the work of the individual class members when they begin to collect the data required. One team is responsible for preparing and standardizing the thiosulfate solution. A second team is responsible for filling the water baths and maintaining their temperature at roughly 5°C above room temperature and at roughly 5°C below room temperature. A third team is responsible for preparing work stations with burets containing each of the solutions needed in the reaction mixture. A fourth team is responsible for assembling the special equipment (bottles for storing solutions, timers, thermometers, etc.) needed for the performance of the experiment.

Procedure

Preparation and Standardization of the Thiosulfate Solution

Weigh out 2.4 g of $Na_2S_2O_3 \cdot 5H_2O$ and dissolve it in approximately 2 L of distilled water. Mix well. Rinse a buret with two small volumes of distilled water, followed by two small volumes of your thiosulfate solution. Then fill the buret with the solution. Record the initial buret reading.

Record the molarity of the standard potassium iodate (KIO_3) solution provided for your use. Rinse a pipet with two small volumes of distilled water, followed by two small volumes of the standard KIO_3 solution. Then use the pipet to deliver 1.00 mL of the standard iodate solution into a clean, 250-mL Erlenmeyer flask. Add 20 mL of distilled water and 20 mL of 0.04 M potassium iodide (KI) to the flask. Swirl the flask to mix the solutions; continue swirling the flask while adding 2 mL of 0.5 M H_2SO_4. The mixture in the Erlenmeyer flask should take on the deep red-brown color characteristic of I_3^- ions.

Procedure in a Nutshell

Prepare a set of solutions containing specified amounts of iron(III) ions, nitric acid, and water. Prepare another set of mixtures containing specified amounts of iodide ions, thiosulfate ions, starch, and water. Mix solutions from the two sets and note the time at which the mixture turns blue, signifying the beginning of a build up of triiodide ions as the thiosulfate ions are consumed. Relate the observed time to the rate of the reaction between iron(III) ions and iodide ions to produce iron(II) ions and triiodide ions. Plot log rate versus log(initial concentration of iodide ions) or log(average concentration of iron(III) ions) to find the values of *a* and *b*, the orders of the reaction with respect to $[Fe^{3+}]$ and $[I^-]$, respectively. Study one mixture at different temperatures to determine the activation energy of the reaction.

Titrate the I_3^- ions with $S_2O_3^{2-}$, adding 1-mL increments of thiosulfate initially, but reducing the incremental size as the color of the titration mixture turns from brown to yellow. Then add 5 mL of 0.2% starch solution. Consult your instructor if the titration mixture does not develop a blue color after starch has been added and the flask has been swirled. Use your wash bottle to rinse all splattered drops from the walls of the flask into the titration mixture. Resume the addition of thiosulfate, adding titrant by drops until the blue color of the starch–iodine complex disappears completely. Usually, only a few additional drops are required. Record the final reading of the thiosulfate buret and calculate the molarity of your thiosulfate solution. Repeat the procedure. Report the molarity obtained in each trial and the average molarity.

Give the thiosulfate solution to the team that is setting up the solution stations. Report the concentration of the thiosulfate solution to the class. When the work stations and water baths are set up, individual students, or pairs of students, proceed to the determination of the orders of the reaction and the value of the activation energy.

Maintaining Constant Temperature

It is inevitable that water that has been cooled below room temperature (by the addition of small amounts of ice that melt completely) will absorb heat from the air and begin to warm up. It is also inevitable that water that has been heated a bit above room temperature will begin to lose heat to the surroundings once heating has stopped. Heat transfer with the surroundings can be minimized by insulating the water baths. However, because the reaction rate is very temperature-sensitive, constant vigilance is required to maintain fairly constant bath temperatures. Team members should be assigned to check the bath temperature regularly and should also be prepared to add ice or hot water to the baths as necessary.

Assembling the Work Stations

The number of work stations needed will depend on the size of the class. There should be at least one station for every six pairs of students. Refilling is easy if automatic burets are used. If these are not available, be sure there is a good supply of each reagent left at the station so that burets can be refilled as necessary. The starch solution should be dispensed from a graduated cylinder; all other solutions can be dispensed from burets. The solutions needed are identified in Table 25.1.

Table 25.1 Reaction Mixtures

	Container 1			*Container 2*			
Trial #	0.04 M Fe^{3+} mL	0.15 M HNO_3 mL	H_2O mL	0.04 M KI mL	0.004 M $S_2O_3^{2-}$ mL	Starch mL	H_2O mL
1	10.00	10.00	30.00	15.00	10.00	5.00	20.00
2	15.00	15.00	20.00	15.00	10.00	5.00	20.00
3	20.00	20.00	10.00	15.00	10.00	5.00	20.00
4	25.00	25.00	0.00	15.00	10.00	5.00	20.00
5	10.00	10.00	30.00	12.00	10.00	5.00	23.00
6	10.00	10.00	30.00	18.00	10.00	5.00	17.00
7	10.00	10.00	30.00	21.00	10.00	5.00	14.00

Glassware

Solutions will need to be kept in a water bath of appropriate temperature for 5–10 minutes before they are mixed. Large beakers and flasks containing small volumes of liquid tend to tip over easily. Your instructor will tell you what measures your class will take to avoid this problem. (For instance, vessels might be clamped in place or they might be weighted down in some manner.)

Obtaining Data on the Kinetics of the Reaction

Part A. Reaction Order with Respect to Fe^{3+}

Prepare the mixture for Trial #1 by adding the solutions specified in Table 25.1 to appropriate containers. Briefly swirl the contents of each container and place the containers in a constant-temperature water bath set at room temperature. Allow a few minutes for the mixtures to reach temperature equilibrium. Meanwhile, prepare the solutions for Trial #2 in a second set of containers and place these also in the water bath. By the time you do that, the solution for Trial #1 should be at bath temperature.

Measure and record the temperature of the Trial #1 solutions. Then simultaneously start the timer and rapidly add the contents of container 1 to container 2. You may temporarily remove the solutions from the water bath when you are doing this. Swirl the solutions until they are well mixed, then return container 2 to the bath. Stop the timer at the first appearance of the blue color. Record the time, *t*, and the temperature (to within 0.1°C). Clean and dry the containers, place the solutions for Trial #3 in them, and return the containers to the water bath. Measure the temperature of the solutions for Trial #2, mix the solutions, and time the reaction as before. Continue in this manner through Trial #4.

Part B. Reaction Order with Respect to I^-

Repeat the procedures specified in Part A, but use the various reaction mixtures given in Table 25.1 for Trials 5, 6, and 7.

Part C. Effect of Temperature

Perform Trials 8, 9, and 10 by preparing reaction mixtures as in Trial #1 and measuring the reaction times in water baths at various temperatures. Trial #8 should be performed at room temperature, Trial #9 at about 5°C higher, and Trial #10 at a temperature about 5°C below room temperature.

Calculations

Using the experimental reaction times and the known concentrations of the reagents, calculate the initial rate, $\frac{1}{2}[S_2O_3^{2-}]_i/\Delta t$, for each experiment.

Calculate the initial concentrations of Fe^{3+} and I^- present in each experiment. Evaluate $[Fe^{3+}]_{av}$ for each experiment, subtracting $[S_2O_3^{2-}]$ from $[Fe^{3+}]_i$ to obtain $[Fe^{3+}]_f$.

Prepare a table showing the following: the initial rate and average Fe^{3+} concentrations for each experiment, the initial I^- concentrations, the initial rate, $\log[Fe^{3+}]_{av}$, $\log[I^-]$, and log(rate).

Using the data for Trials 1 through 4, plot log(rate) versus $\log[Fe^{3+}]_{av}$. Draw the best straight line through the experimental points, and calculate its slope. This is equal to a, the reaction order for Fe^{3+}.

Using the data from Trials 1, 5, 6, and 7, plot log(rate) versus $\log[I^-]$. Draw the best straight line and calculate its slope. This is equal to b, the reaction order I^-.

Round off a and b to integers and write the resulting rate expression. Calculate the initial rates for Trials 8, 9, and 10. Substitute the measured rates, $[Fe^{3+}]_{av}$ and $[I^-]_i$ in the rate expression to obtain the specific rate constant, k, at each of the three temperatures. Prepare a table showing k, $\ln k$, T, and $1/T$. Plot $\ln k$ versus $1/T$. Measure the slope and from it calculate the activation energy, E_a.

Disposal of Reagents

The acidic solutions (HNO_3 and $Fe(NO_3)_3$) should be neutralized with dilute NaOH solution. The neutralized chemicals may be diluted and poured down the drain. All other solutions may also be diluted and poured down the drain.

Questions

1. For reactions in aqueous solution, would it be easy or difficult to determine the order with respect to H_2O? Explain.

2. Why is it necessary to convert $[Fe^{3+}]_i$ to $[Fe^{3+}]_{av}$? Why is it unnecessary to convert $[I^-]_i$ to $[I^-]_{av}$?

name | section | date

Pre-Lab Exercises for Experiment 25

These exercises are to be completed after you have read the experiment but before you come to the laboratory to perform it.

1. The reaction between bromate ions and bromide ions occurs according to the equation shown below.

$$BrO_3^-(aq) + 5Br^-(aq) + 6H^+(aq) \rightarrow 3Br_2(l) + 3H_2O(l)$$

The following table shows the results of four experiments done on this system. Use these data to determine the order of reaction with respect to each reactant, the overall order of reaction, and the value of the rate constant.

Trial #	$[BrO_3^-]_i$	$[Br^-]_i$	$[H^+]_i$	*Measured initial rate* $-\Delta[BrO_3^-]/\Delta t$, (mol/L·s)
1	0.10	0.10	0.10	7.9×10^{-4}
2	0.20	0.10	0.10	1.7×10^{-3}
3	0.20	0.20	0.10	3.1×10^{-3}
4	0.10	0.10	0.20	3.2×10^{-3}

2. Methane gas reacts with diatomic sulfur in the gas phase, producing carbon disulfide and hydrogen sulfide. The rate constant for this reaction is 1.1 L/mol·s at 550°C. It is 6.4 L/mol·s at 625°C. Use these data to determine the value of the activation energy for the reaction. When *k* is measured at only two temperatures, Equation 25.13 may be written as shown below.

$$\ln k_2 - \ln k_1 = -\frac{E_a}{R}\left(\frac{1}{T_2} - \frac{1}{T_1}\right)$$

name	section	date

Summary Report on Experiment 25

Standardization of the $Na_2S_2O_3$ *Solution*

	Trial 1	*Trial 2*	*Trial 3*[*]
Final buret reading, KIO_3			
Initial buret reading, KIO_3			
Volume of KIO_3 delivered			
Final buret reading, $Na_2S_2O_3$			
Initial buret reading, $Na_2S_2O_3$			
Volume of $Na_2S_2O_3$ used			
Molarity of $Na_2S_2O_3$			
Average molarity of $Na_2S_2O_3$			

Trial #	*Initial Temperature*	*Reaction Time* Δt*, sec*
1		
2		
3		
4		
5		
6		
7		
8		
9		
10		

[*] Optional

Results of Trials 1 through 4

$[S_2O_3^{2-}]_i$ = ________ **$[I^-]_i$ = ________**

$\log[I^-]_i$ = ________

Trial #	$[Fe^{3+}]_i$	$[Fe^{3+}]_f$	$[Fe^{3+}]_{av}$	$\log[Fe^{3+}]_{av}$
1	________	________	________	________
2	________	________	________	________
3	________	________	________	________
4	________	________	________	________

Trial #	*Time*	*Initial Rate*	$\log\left(\text{Initial Rate}\right)$
1	________	________	________
2	________	________	________
3	________	________	________
4	________	________	________

Procedure for calculating *a*, the reaction order with respect to Fe^{3+}:

name ______ section ______ date ______

Results of Trials 5 through 7

$[Fe^{3+}]_i$ = ______ $[S_2O_3^{2-}]_i$ = ______ $[Fe^{3+}]_f$ = ______

$[Fe^{3+}]_{av}$ = ______

$\log[Fe^{3+}]_{av}$ = ______

Trial #	$[I^-]_i$	$\log[I^-]_i$	*Time*	*Initial Rate*	$\log\left(\text{Initial Rate}\right)$
5	______	______	______	______	______
6	______	______	______	______	______
7	______	______	______	______	______

Procedure for calculating *b*, the reaction order with respect to I^-:

Experimental rate expression for the reaction:

Results of Trials 8 through 10

$[Fe^{3+}]_{av}$ = ________

$[I^-]_i$ = ________

Trial #	*Time*	*Rate*	*k*	ln *k*	*T*	1/*T*
8	________	________	________	________	________	________
9	________	________	________	________	________	________
10	________	________	________	________	________	________

Procedure used to calculate *k*:

Experimental value for the activation energy E_a of the reaction: ________

Procedure used to calculate E_a:

EXP **26**

Qualitative Analysis of Household Chemicals

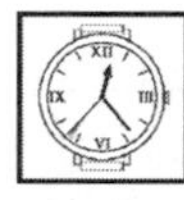

Laboratory Time Required Three hours.

Special Equipment and Supplies

Balance
Thermometer
Conductivity tester
Labels

Unknowns
Photographic fixer
Cornstarch
Chalk
Baking soda
Washing soda
Table salt
Sugar
Epsom salt
Alum
Sour salt

Reagents
Acetic acid, $CH_3COOH(aq)$
Ammonia, $NH_3(aq)$
Iodine, $I_2(alcohol)$
Red cabbage juice
Rock salt
Ice

Objective

In performing this experiment, students will become familiar with the chemical properties of many household materials.

Safety

Some institutions require that safety glasses or goggles must be worn whenever work is being done in a laboratory environment. Do not remove any of the unknowns or reagents from the laboratory for home use.

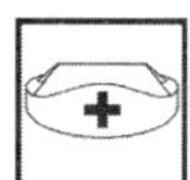

First Aid

For exposure to any of the materials used in this experiment, thoroughly flush the skin or eyes with water. Chemicals may be removed from the stomach by drinking large amounts of water and inducing vomiting.

Preamble

Several years ago, the American Chemical Society offered a prize to anyone who could find a substance that was not a chemical. The point, of course, was that all matter, whether "natural" or "artificial," is composed of atoms and molecules. Although most college students accept this truism readily, they still often view the chemicals they work with in the laboratory as something different from the materials that abound in the "real world." This experiment, in which the "active ingredient" in many household products will be identified, is an attempt to bridge the gap between the chemistry lab and the household environment.

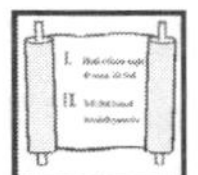

Principles

You will be provided with a number of white solids, identified only by an alphanumeric code. Your job will be to identify each of these unknowns correctly. The possibilities are table salt, sugar, epsom salt, alum, photographic fixer, cornstarch, sour salt, chalk, baking soda, and washing soda. Many of these materials are likely to be found in the typical home. Others, while not as common, are readily available in pharmacies, building supply stores, or camera shops, and may be purchased without any special license or prescription.

Suggestions for identifying the unknowns using chemical tests are made below. Note that many of the test reagents are themselves ingredients in products commonly used in the home. For instance, acetic acid (CH_3COOH or $HC_2H_3O_2$) is a weak acid, with a characteristic sour taste; vinegar is essentially a 1 M solution of acetic acid. Ammonia is a gas in pure form; its aqueous solutions are alkaline and are commonly used as household cleansers. Pure iodine is a solid that sublimes (passes directly into the gas phase) at slightly elevated temperatures. Alcoholic solutions of iodine (tinctures of iodine) are often used as disinfectants. Red cabbage juice, obtained by boiling red cabbage leaves in water, is a natural indicator that displays a variety of colors in the presence of substances that behave as acids or bases of various strengths.

The information given below should enable you to devise a plan for identifying the unknowns. The formulas given are those of the main ingredient in each household item. (Of course, the materials used in the home are generally not pure and there is no single formula for the complete mixture.)

Insoluble in Water

Chalk (main ingredient: $CaCO_3$, calcium carbonate) and cornstarch (main ingredient: a polymer of glucose, $(C_6H_{10}O_5)_X$, are insoluble in water. Like all carbonates, chalk will fizz when treated with acid. Starch will turn blue when treated with iodine.

Soluble in Water

Form Precipitates When Treated with Ammonia

The unknowns other than chalk and cornstarch are soluble in water. Epsom salt (main ingredient: $MgSO_4 \cdot 7H_2O$, magnesium sulfate heptahydrate) dissolves with a noticeable cooling effect. Addition of aqueous ammonia to a solution of epsom salt produces "milk of magnesia," a suspension of magnesium hydroxide that has a milky appearance.

Alum (main ingredient: $NH_4Al(SO_4)_2 \cdot 12H_2O$, ammonium aluminum sulfate dodecahydrate) is used as an astringent and as a pickling agent. Addition of aqueous ammonia to a solution of alum produces a gelatinous precipitate of aluminum hydroxide. The other water-soluble unknowns do not form precipitates when treated with ammonia.

No Precipitate When Treated with Ammonia

Washing soda (main ingredient: Na_2CO_3, sodium carbonate) dissolves in water to produce solutions basic enough to impart a lime green color to red cabbage juice. Baking soda (main ingredient: $NaHCO_3$, sodium hydrogen carbonate) also has basic properties; adding baking soda to water that contains red cabbage juice will change the color of the solution to blue-green (aqua). Both washing soda and baking soda, like other carbonates, will fizz when treated with acetic acid.

Sour salt (main ingredient: $C_6H_8O_7$, citric acid) is often used in the preparation of "sweet and sour" meats. Red cabbage juice acquires an orange-pink (rose) color upon the addition of citric acid. Adding sour salt to solutions of washing soda may also cause fizzing.

Photographic fixer (main ingredient: $Na_2S_2O_3 \cdot 5H_2O$, sodium thiosulfate pentahydrate) is soluble in water and is capable of reducing I_2 molecules to I^- ions. It will decolorize brown iodine solutions and also remove the blue color of the iodine-starch complex.

No Reaction with Test Reagents

Table salt (main ingredient: NaCl, sodium chloride), like epsom salt, alum, fixer, washing soda, and baking soda, produces ions when dissolved in water. Thus, solutions of table salt will conduct electricity. These solutions also have very low freezing points. A solution of 4 g of table salt in 10 mL of water will not freeze in an ice/rock salt/water bath (bath temperature: $-10°C$) but a solution of 4 g of table sugar (main ingredient: $C_{12}H_{22}O_{11}$, sucrose) in 10 mL of water will freeze when placed in such a bath. In addition, sugar molecules do not dissociate in aqueous solution, so solutions of sucrose do not conduct electricity. Neither sugar nor salt solutions react with acetic acid, ammonia, red cabbage juice, or iodine.

Procedure

Devise a scheme for separating the unknowns into groups and then identifying the members of each group. You may, if you wish, base your procedure on the four groups discussed in the Principles section or you may try an alternate approach, such as trying to separate the electrolytes (those whose solutions conduct electricity) from the nonelectrolytes (those whose solutions do not conduct electricity) or you may try separating those materials that fizz when treated with acid from those that do not. No matter where you start, you will find it helpful to follow the suggestions given below.

Procedure in a Nutshell

Use the information given in the Principles section to devise a scheme for identifying each of the unknowns.

1. In determining solubility, use only a pea-sized (or smaller) quantity of solid, placed in a test tube. Add roughly 10 mL of water and mix well. You need just enough solid to be seen easily if it does not dissolve, but not enough to form a saturated solution if it is moderately soluble. A slight cloudiness may be due to a trace of insoluble filler and should not lead to a conclusion of "insoluble" if the major portion of the sample dissolves.

2. Do not add anything to the containers of unknowns or test reagents. Do not use spatulas from one container in another.

3. To determine if an unknown will fizz when treated with acid, place a pea-sized quantity of solid in a test tube and add a few drops of acetic acid.

4. When testing materials with iodine, add only 2–3 drops of reagent. When testing the effects of materials on the color of red cabbage juice, place two or three drops of juice in a few milliliters of deionized water (enough to give the water a pale pink color); then add a pea-sized quantity of solid to the water containing the red cabbage juice.

5. Remember that reagents may interfere with one another. For instance, if you have several solutions and add iodine to each, all of the solutions, except the one containing the photographic fixer, will turn brown. Adding red cabbage juice to the brown solutions would then be futile; the brown color would interfere with your being able to observe the characteristic color that the juice acquires when its acidity is changed. Use fresh samples whenever it seems necessary.

6. Your instructor will show you how to test solutions for conductivity.

7. Be sure to record the identity code of each unknown and any data collected that would help you to identify the unknown.

Disposal of Chemicals

The household chemicals that are insoluble in water may be discarded in the containers designated for solid, nonhazardous waste. All soluble household chemicals and test reagents may be diluted and flushed down the sink with a voluminous stream of water.

name | section | date

Pre-Lab Exercises for Experiment 26

This exercise is to be completed after you have read the experiment but before you come to the laboratory to perform it.

1. Complete the table shown below. The first line is completed as a guide.

	Soluble in H_2O?	*Solution conducts electricity?*	*Reacts with vinegar?*	*Reacts with red cabbage juice?*	*Precipitate with NH_3?*	*Solution of 4 g/10 mL freezes?*	*Reacts with I_2?*
Epsom salt	yes	yes	no	no	yes; milky	yes	no
Sugar							
Table salt							
Alum							
Photographic fixer							
Cornstarch							
Chalk							
Baking soda							
Washing soda							
Sour salt							

2. Devise an analysis scheme that will permit you to identify each of the unknowns.

name ______ section ______ date ______

Summary Report on Experiment 26

Unknown label ______

Material Tested	*Test Reagent*	*Procedure or Observation*

Unknown label ______

Material Tested	*Test Reagent*	*Procedure or Observation*

Unknown label ______

Material Tested	*Test Reagent*	*Procedure or Observation*

Unknown label ______

Material Tested	*Test Reagent*	*Procedure or Observation*

Unknown label ______________________

Material Tested	*Test Reagent*	*Procedure or Observation*

Unknown label ______________________

Material Tested	*Test Reagent*	*Procedure or Observation*

Unknown label ______________________

Material Tested	*Test Reagent*	*Procedure or Observation*

Unknown label ______________________

Material Tested	*Test Reagent*	*Procedure or Observation*

name section date

Unknown label

Material Tested | *Test Reagent* | *Procedure or Observation*

Unknown label

Material Tested | *Test Reagent* | *Procedure or Observation*

Use this table to summarize your data.

Unknown	*Soluble in H_2O?*	*Solution conducts electricity?*	*Reacts with vinegar?*	*Reacts with red cabbage juice?*	*Precipitate with NH_3?*	*Solution of 4 g/10 mL freezes?*	*Reacts with I_2?*	*Identity of unknown*

Analysis of a Soluble Salt

EXP
27

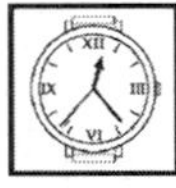

Laboratory Time Required

One to three hours, depending on the number of salts to be analyzed.

Special Equipment and Supplies

Test wires
Microspatulas
Centrifuges
Litmus paper (red and blue)
Steel paper clips
Sandpaper
Cobalt blue glass

6 M ammonia, NH_3
6 M nitric acid, HNO_3
6 M hydrochloric acid, HCl
6 M sodium hydroxide, NaOH
0.05 M silver nitrate, $AgNO_3$
1 M barium chloride, $BaCl_2$
3 M acetic acid, CH_3COOH
Dimethylglyoxime, DMG (*s*)
Sodium chloride, NaCl (*s*)
Unknown salts

Objective

In performing this experiment, the student will become familiar with the behavior of some cations and anions and determine the identities of various pure salts via qualitative analysis.

Safety

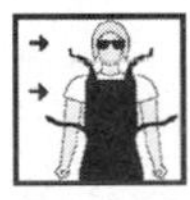

Avoid getting acids or bases in your eyes or on your skin.

When centrifuging mixtures, always use pairs of test tubes to help keep the centrifuge in balance. Wait for the centrifuge to come to a complete stop, after it is turned off, before reaching in to get your test tubes out. Never touch a spinning centrifuge.

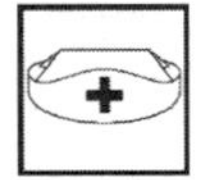

First Aid

Following skin contact with any of the reagents, wash the area with copious amounts of water.

If acid or base enters your eyes, use the eyewash fountain to flush the chemical away, then consult a doctor.

Preamble

In the past, qualitative analysis formed a major part of the laboratory program in introductory chemistry courses. A number of factors have led to a decrease in the amount of the time spent on qualitative analysis — concern over the toxicity of reagents, development of alternative methods of analysis (such as atomic absorption spectroscopy), a perception that students did not understand the complicated chemistry needed to separate the various ions, and so on. This experiment is a simple one, from which the need to perform separations has been eliminated. Nevertheless, it illustrates many interesting tests and introduces students to the analysis of anions, an area often ignored in traditional qualitative analysis.

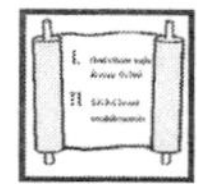

Principles

Suppose someone handed you a vial containing a crystalline solid and said, "You're taking chemistry. Find out what's in this vial, please." Although you might like to oblige the petitioner, you'd have to admit that the request is fairly unreasonable, even impossible.

Yet, in this experiment, you will indeed be asked to identify the contents of several vials. However, the problem will be made more tractable by the formulation of some ground rules. All solids you will be asked to identify will be pure salts that are soluble in water. Each will contain only one type of cation and one type of anion. In fact, only six positive ions and five negative ions will be represented.

Procedure

Procedure in a Nutshell

Obtain an unknown. Follow the first set of instructions to determine which cation is present in the unknown. Then, follow the next set of instructions to determine which anion is present as well.

Obtain an unknown and record its identification code on the Summary Report Sheet. Perform the cation analyses described below to determine whether your unknown is a salt of NH_4^+, Ba^{2+}, Cu^{2+}, Ni^{2+}, K^+, or Na^+. Then perform the anion analyses to determine whether your unknown contains CO_3^{2-}, Cl^-, CrO_4^{2-}, I^-, or SO_4^{2-} ions. Once you have identified both the anion and cation in one salt, repeat the process with another unknown. Try to do as many analyses as time permits. Prepare a new Summary Report Sheet for each unknown. Use the Summary Report Sheet provided as a guide for your extra report forms.

Use your data and the supplementary information found in Table 27.1 to identify the cation and anion present in each salt you analyze. Use the equations given in Table 27.2 as a guide in writing balanced net ionic equations for the tests involved in making each identification. Give the correct chemical formula for each salt you analyze.

Cation Analyses

Dissolve a spatula-tipful of unknown in 5 mL of distilled water. Stir the mixture. If the solid does not dissolve completely, consult your instructor.

Pour one-half of the solution containing your unknown into a small beaker. Cautiously add 6 M NaOH, until the solution is just basic. Then add an additional 3 drops of sodium hydroxide. Observe the contents of the beaker. If a precipitate has formed, your unknown is a barium, copper(II), or nickel(II) salt. Proceed to the tests for Ba^{2+}, Cu^{2+}, and Ni^{2+} described below.

If no precipitate has formed, your unknown is a salt of NH_4^+, K^+, or Na^+. To determine if your unknown is an ammonium salt, proceed as follows. Moisten a strip of red litmus paper and place it securely on the convex side of a watch glass (see Figure 7.2). Place the watch glass on top of the beaker that contains the unknown solution and NaOH. If the color of the litmus paper has not changed after 2–3 minutes, heat the beaker gently. If the litmus paper turns completely blue, your unknown contains the ammonium ion. Proceed to the anion tests.

If the test for NH_4^+ is negative, do a flame test on the portion of your unknown solution to which NaOH has not been added. Clean a test wire by alternately dipping it in 6 M HCl and heating it in the burner flame. When the wire does not impart any color to the flame, dip it into the unknown solution. Be sure a thin layer of solution fills the test wire loop and bring the loop to the edge of the outer cone of the burner flame, at the height of the inner cone. If a yellow-orange flame is observed, your unknown is a sodium salt. If a violet flame is seen, your unknown is a potassium salt. (Viewing the flame through a piece of cobalt glass may aid in the detection of the violet flame.) Proceed to the anion analyses.

To test for the presence of Ba^{2+}, Cu^{2+}, or Ni^{2+}, decant the liquid from the beaker containing the precipitate that formed when NaOH was added to the unknown solution. While mixing, add 5 mL of 3 M NH_3. If the precipitate dissolves, producing a royal-blue solution, your unknown is probably a salt of copper(II). Confirm that the cation is the copper(II) ion by adding 3 M $HC_2H_3O_2$, dropwise, to the royal-blue solution until it is nearly colorless. Then add a few grains of NaCl. Gently sand a steel paper clip and place it in the solution. Leave the paper clip in the solution, undisturbed, for 15–30 minutes. If a uniform brown coating forms on the paper clip, the cation is definitely copper(II). Proceed to the anion analyses.

If the hydroxide precipitate dissolves to give a green solution, your unknown is probably a salt of nickel(II). Confirm that the cation is nickel(II) by adding 3 M $HC_2H_3O_2$, by drops, until the solution is nearly colorless. Then, add a few grains of DMG. If a red precipitate is produced, the cation is nickel(II). Proceed to the anion analyses.

If the hydroxide precipitate does not dissolve in ammonia, dissolve it in 6 M hydrochloric acid. Then add 3 drops of 6 M H_2SO_4. If a white precipitate forms, the cation is probably Ba^{2+}. Confirm the presence of Ba^{2+} in your unknown with a flame test.

Centrifuge the test tube containing the probable $BaSO_4$ precipitate. Add 1 drop more of 3 M H_2SO_4 to test for complete precipitation; centrifuge the tube once again. Continue centrifuging and testing the centrifugate until all Ba^{2+} has been removed from the solution.

Clean a test wire by dipping it in concentrated HCl and heating it in the burner flame. Repeat the process of dipping and heating until the wire does not impart any color to the flame. Decant the centrifugate from the test tube containing the probable $BaSO_4$ precipitate. Add 5 drops of concentrated HCl to the precipitate. Stir the mixture of $BaSO_4$ and HCl and dip the test wire into the resulting suspension. Bring the loop of the test wire to the edge of the outer cone of the burner flame, at the height of the top of the inner cone. The appearance of a green flame indicates that the precipitate is barium sulfate. Proceed to the anion analyses.

Anion Analyses

Dissolve a spatula-tipful of unknown in 5 mL of distilled water. Stir the mixture. Pour roughly equal portions of solution into each of three test tubes.

Add 6 M HNO_3 to the first of the three test tubes, by drops, while stirring, until the solution is just acid to litmus. If your solution was originally yellow and it turns orange upon addition of acid, your unknown is a chromate salt. If bubbles form or fizzing is heard, your salt is a carbonate. Proceed to another unknown.

If your solution did not turn orange or fizz when acid was added, add 10 drops of 1 M $BaCl_2$ to the second test tube. If a white precipitate forms, centrifuge the tube, using a tube filled with distilled water as a counterbalance. Decant the centrifugate (the liquid portion) and discard it. Treat the precipitate with several drops of 6 M HNO_3. If the precipitate fizzes and dissolves, your unknown is a carbonate. If the precipitate does not dissolve, your unknown is a sulfate. Proceed to another unknown.

If your salt is not a chromate, carbonate, or sulfate, add 10 drops of 0.05 M $AgNO_3$ to the third test tube. If a white precipitate forms, your unknown is a chloride. If a yellow precipitate forms, your unknown is an iodide. If you cannot determine the color of the precipitate, centrifuge the tube and discard the centrifugate. Add 1 mL of 6 M NH_3 to the precipitate, while stirring. Silver chloride will dissolve, silver iodide will not. Proceed to another unknown.

Disposal of Reagents

Any solutions containing silver or nickel(II) ions should be placed in labeled collection bottles. All other reagents can be neutralized, diluted, and flushed down the drain.

Table 27.1 Solubility Rules

1. All common sodium, potassium, and ammonium salts are soluble in water.
2. The chloride and iodide salts of all common metals, except silver, lead, and mercury(I), are soluble in water. Lead chloride is soluble in hot water.
3. The sulfates of all metals, except lead, mercury(I), barium, strontium, and calcium are soluble in water.
4. The carbonates of all metals, except those of Group IA and ammonium carbonate, are insoluble in water.
5. Most metal hydroxides are insoluble in water. However, the hydroxides of Group IA are soluble and those of Group IIA are moderately soluble.

Table 27.2 Reactions

Ammonium	$NH_4^+ + OH^- \longrightarrow NH_3\uparrow + H_2O$
	$NH_3 + \underset{\text{red}}{HIn} \longrightarrow \underset{\text{blue}}{NH_4^+ + In^-}$
Barium	$Ba^{2+} + SO_4^{2-} \longrightarrow BaSO_4\downarrow$
Carbonate	$CO_3^{2-} + 2H^+ \longrightarrow H_2O + CO_2\uparrow$
	$Ba^{2+} + CO_3^{2-} \longrightarrow BaCO_3\downarrow$
	$BaCO_3 + 2H^+ \longrightarrow H_2O + CO_2\uparrow + Ba^{2+}$
Chromate	$CrO_4^{2-} + 2H^+ \longrightarrow Cr_2O_7^{2-} + H_2O$
Chloride	$Ag^+ + Cl^- \longrightarrow AgCl\downarrow$
	$AgCl + 2NH_3 \longrightarrow Ag(NH_3)_2^+ + Cl^-$
Copper	$Cu^{2+} + 2OH^- \longrightarrow Cu(OH)_2\downarrow$
	$Cu(OH)_2 + 4NH_3 \longrightarrow Cu(NH_3)_4^{2+} + 2OH^-$
	$Cu(NH_3)_4^{2+} + 2HC_2H_3O_2 \longrightarrow Cu^{2+} + 2NH_4C_2H_3O_2$
	$Cu^{2+} + Fe \longrightarrow Cu + Fe^{2+}$
Iodide	$Ag^+ + I^- \longrightarrow AgI\downarrow$
Nickel	$Ni^{2+} + 2OH^- \longrightarrow Ni(OH)_2$
	$Ni(OH)_2 + 6NH_3 \longrightarrow Ni(NH_3)_6^{2+} + 2OH^-$
	$Ni(NH_3)_6^{2+} + 2HC_2H_3O_2 \longrightarrow Ni^{2+} + 2NH_4C_2H_3O_2$
	$Ni^{2+} + 2DMG \longrightarrow Ni(DMG)_2\downarrow$
Sulfate	$Ba^{2+} + SO_4^{2-} \longrightarrow BaSO_4\downarrow$
	$BaSO_4 + H^+ \longrightarrow NR$

Questions

1. Glenn Student dissolved a small amount of unknown salt in water and poured one-half of the resulting solution into a beaker. Next, Glenn made the solution basic with NaOH. No precipitate formed, so Glenn dipped a piece of red litmus paper into the solution. When the litmus paper turned blue, Glenn concluded the unknown was an ammonium salt. Why is Glenn's method for finding ammonium inconclusive? What, if anything, can you say about the identity of Glenn's salt?

2. Chris Frosh found that an unknown was a barium salt. Chris concluded that the unknown could not be a carbonate or sulfate. Do you concur with Chris? Explain your reasoning.

name section date

Pre-Lab Exercises for Experiment 27

These exercises are to be performed after you have read the experiment but before you come to the laboratory to perform it.

1. Name a single reagent and/or physical property that could be used to distinguish between:
   a. $K_2CrO_4(s)$ and $KCl(s)$

   b. $BaCl_2(s)$ and $BaCO_3(s)$

   c. $KI(aq)$ and $NaCl(aq)$

   d. $Cu(OH)_2(s)$ and $KaOH(s)$

2. A white, crystalline solid dissolved in water forming a pale blue solution. A precipitate resulted when the solution was treated with 6 M sodium hydroxide. The precipitate dissolved in ammonia, giving a royal-blue solution. A second portion of the solid was dissolved in water. Treatment of a portion of that solution with silver nitrate resulted in a precipitate. The precipitate did not dissolve in ammonia. Identify the original salt, giving its chemical formula. Briefly explain your reasoning.

name | section | date

Summary Report on Experiment 27

Unknown identification code ______________________

Description of unknown ______________________

Cation Analysis

Material Tested	*Reagent*	*Observation*	*Equation*

Identity of cation ______________________

Anion Analysis

Material Tested	*Reagent*	*Observation*	*Equation*

Identity of anion ____________________

Formula of salt ____________________

name _______________ section _______________ date _______________

Unknown identification code _______________

Description of unknown _______________

Cation Analysis

Material Tested	*Reagent*	*Observation*	*Equation*

Identity of cation _______________

Anion Analysis

Material Tested	*Reagent*	*Observation*	*Equation*

Identity of anion ______________________

Formula of salt ______________________

name   section   date

Unknown identification code

Description of unknown

Cation Analysis

Material Tested	*Reagent*	*Observation*	*Equation*

Identity of cation

Anion Analysis

Material Tested	*Reagent*	*Observation*	*Equation*

Identity of anion ____________________

Formula of salt ____________________

name	section	date

Unknown identification code ____________

Description of unknown ____________

Cation Analysis

Material Tested	*Reagent*	*Observation*	*Equation*

Identity of cation ____________

Anion Analysis

Material Tested	*Reagent*	*Observation*	*Equation*

Identity of anion ______________________

Formula of salt ______________________

name ______________________ section ______________________ date __________

Unknown identification code ______________________

Description of unknown ______________________

Cation Analysis

Material Tested	*Reagent*	*Observation*	*Equation*

Identity of cation ______________________

Anion Analysis

Material Tested	*Reagent*	*Observation*	*Equation*

Identity of anion ______________________

Formula of salt ______________________

Mathematical Operations

APP A

It is most likely that you will use a calculator when you need to work with logarithms and that you will prepare any needed graphs with the help of a computer. Nonetheless, it is useful for you to have some familiarity with the properties of logarithms and with the manual preparation of graphs. This is because calculators occasionally malfunction or give the wrong answer because the data was not keyed in properly. In addition, it is often helpful to do rough graphs in the lab (to see if additional data is needed or to see if a data point seems to be in error and should be rechecked).

Logarithms

Common Logarithms

The logarithm X of a number N is defined by the relations

$$\log_a N = X \tag{A.1}$$

where

$$N = a^x \tag{A.2}$$

For $a = 10$, the common logarithm is obtained, while for a = 2.71828 (represented by the symbol e), the natural logarithm is obtained. Common logarithms are generally used in the mathematical operations of multiplication, division, and finding powers and roots. Natural logarithms are used in certain thermodynamic relations, distribution functions, and so on. In this laboratory manual the common logarithm of a number is represented by $\log N$; the natural logarithm is represented by $\ln N$.

Most calculators have keys for both $\log N$ and $\ln N$. If your calculator has only one of these functions, you may use Equation A.3 to convert values from common to natural logarithms, or vice versa.

$$\ln N = 2.303 \log N \tag{A.3}$$

In the days before inexpensive calculators were available, logarithms were frequently used to simplify difficult calculations. The relations shown below reveal the simplifying power of logarithms.

$$\log XY = \log X + \log Y \tag{A.4}$$

$$\log (X/Y) = \log X - \log Y \tag{A.5}$$

$$\log Y^X = X \log Y \tag{A.6}$$

To find the common logarithm of a number *N*, it is convenient to write *N* as an exponential number:

$$N = M \times 10^Z \tag{A.7}$$

where *Z* is an integer and *M* is a number between 1 and 10. For example, 456 would be written 4.56×10^2, while 0.0456 would be written as 4.56×10^{-2}. The logarithm of *N* is then given by Equation A.8.

$$\log N = \log M + Z \tag{A.8}$$

for example, $$\log 4.56 \times 10^2 = \log 4.56 + 2$$

and $$\log 4.56 \times 10^{-2} = \log 4.56 - 2$$

If this convention is adopted, it is obviously necessary to tabulate only the logarithms of numbers between 1 and 10 inclusive. Such a tabulation is given in Table A.1. The logarithm of a two-digit number may be read directly from the table; the logarithm of a three-digit number may be estimated by interpolation.

Thus, the logarithm of 4.56 lies between the logarithm of 4.5 (0.653) and that of 4.6 (0.663). Because six-tenths of the difference between 0.663 and 0.653 is easily determined to be 0.006, the logarithm of 4.56 is 0.653 plus 0.006, or 0.659. By using Equation A.8, we find the logarithm of 4.56×10^2 is 0.659 + 2, or 2.659. Similarly, the logarithm of 4.56×10^{-2} is $0.659 - 2$ or -1.341.

By reversing the process, any number whose logarithm is given may be determined. Such a number is called the antilog, or $\log^{-1}$, of the number *X*. The antilog of 2.659 is $10^{2.659}$, which is equal to $10^{0.659} \times 10^2$. Referring to Table A.1, you will find the antilog of 0.659 to be 4.56. Therefore the antilog of 2.659 is 4.56×10^2. Similarly, the antilog of -1.341 is $10^{-1.341}$, which is equal to 4.56×10^{-2}. By using the table, you will find 4.56×10^{-2} to be the antilog of -1.341.

To multiply or divide using logarithms, you use Equations A.4 and A.5. For example, to multiply 2.1 by 3.3, add the logarithms of 2.1 and 3.3.

$$\log 2.1 = 0.322$$

$$\log 3.3 = 0.518$$

$$(\log 2.1 + \log 3.3) = 0.840$$

$$\text{antilog} 0.840 = 6.9$$

$$2.1 \times 3.3 = 6.9$$

To divide 2.1 by 3.3, subtract the logarithm of 3.3 from the logarithm of 2.1.

$$\log 2.1 = 0.322$$

$$-\log 3.3 = -0.518$$

$$(\log 2.1 - \log 3.3) = -0.196 = 0.804 - 1$$

$$\text{antilog } (0.804 - 1) = 6.4 \times 10^{-1}$$

$$(2.1 \div 3.3) = 0.64$$

Note the somewhat unusual third step, in which we added and subtracted 1 from –0.186. This was done to obtain a positive number that could be found in the log table. This step is necessary only when a number is divided by a larger number.

To raise a number to a power using logarithms, use Equation A.6. Note that X may be an integer or a fraction, so that Equation A.6 may be used for "raising to a power" or for "taking roots."

We illustrate this below by finding the value of 2^3 and $\sqrt[3]{8}$ $(= 8^{1/3})$.

$$\log 2^3 = 3\log 2 = 3(0.301) \qquad = 0.903$$

$$\text{antilog } 0.903 \qquad = 8$$

$$2^3 \qquad = 8$$

$$\log 8^{1/3} = (1/3)\log 8 = (1/3)(0.903) \qquad = 0.301$$

$$\text{antilog } 0.301 \qquad = 2$$

$$8^{1/3} \qquad = 2$$

Natural Logarithms

Natural logarithms follow the same combination laws as do common logarithms but are somewhat more difficult to work with because of the change of base from 10 to 2.71828. Although Equations A.4, A.5, and A.6 apply to natural logarithms, it is generally easier to perform simple mathematical operations using common logarithms and to use natural logarithms only when the natural log appears in the equation of interest. The natural logarithm may then be obtained by (1) consulting a table of natural logarithms, (2) using the ln key on your calculator, or (3) calculating the natural log from the common logarithm of the number using Equation A.3.

Table A.1 Three-Place Common Logarithms

N	.0	.1	.2	.3	.4	.5	.6	.7	.8	.9
1	0.000	0.041	0.079	0.114	0.146	0.176	0.204	0.230	0.255	0.279
2	0.301	0.322	0.342	0.362	0.380	0.398	0.415	0.431	0.447	0.462
3	0.477	0.491	0.505	0.518	0.531	0.544	0.556	0.568	0.580	0.591
4	0.602	0.613	0.623	0.633	0.643	0.653	0.663	0.672	0.682	0.690
5	0.699	0.708	0.716	0.724	0.732	0.740	0.748	0.756	0.763	0.771
6	0.778	0.785	0.792	0.799	0.806	0.813	0.820	0.826	0.833	0.839
7	0.845	0.851	0.857	0.863	0.869	0.875	0.881	0.886	0.892	0.898
8	0.903	0.908	0.914	0.919	0.924	0.929	0.934	0.940	0.944	0.949
9	0.954	0.959	0.964	0.968	0.973	0.978	0.982	0.987	0.991	0.996
10	1.000									

Graphs

General Principles

The grid paper in your lab notebook will suffice as graph paper when you are preparing rough plots while you are in lab and taking data. The graphs that are submitted with your reports should be constructed more carefully. Follow the instructions given below whether you are preparing your graphs by hand or you are using a computerized graphing package.

1. If you are preparing your graph manually, use a good grade of graph paper (preferably one with millimeter divisions). If you are planning to read values from a computer-generated graph, set the options to print "minor ticks" between major scale divisions.

2. Clearly specify the variable to be plotted along each axis. Note that your choice of axes is not arbitrary. For instance, if you expect your data to fit the linear equation, $y = mx + b$, you should plot y on the vertical axis and x on the horizontal axis.

3. Mark the scales clearly along each axis. Choose the scale so that the coordinates of any point on the graph may be easily and accurately determined without computation.

4. Choose the scale so that the coordinates of any point may be estimated to the same number of significant figures as the original data. When observance of this rule requires a graph of unreasonably large size, it may be possible to portray all the data on a graph of reasonable size by using a relatively coarse scale. The portion of greatest interest can then be shown using a finer scale, either on a separate graph or as an insert on the main graph.

5. Surround each experimental point by a circle, diamond, or square, so that it is easily recognized.

6. If you are preparing your plots manually, draw straight lines through the experimental points using a ruler. Use a French curve in constructing curved lines. Do not zigzag between points. Note that the best experimental line may not pass through every point.

7. Label the finished graph with a suitable title.

Calculating the Slope

The slope of line *AB* may be easily calculated using the coordinates of any two points that lie on the line. For example, the points P_1 and P_2 having the coordinates (X_1, Y_1) and (X_2, Y_2) may be used. The slope, *m*, of the line *AB* is then given by Equation A.9:

$$m = \frac{Y_2 - Y_1}{X_2 - X_1} \tag{A.9}$$

Note that the coordinates of P_1 and P_2 do not depend on the scale used. Therefore, the slope is independent of the scale used.

Using a Computer to Prepare Graphs

Many computer spreadsheet programs, such as Excel™ and QuadroPro™, include graphing tools. The following is a general guideline representing necessary components for constructing graphs with such computer programs. The specific instructions/tips given below apply directly to Excel™. Of course, you will need directions for your particular software, but they are likely to be quite similar to the methods discussed here.

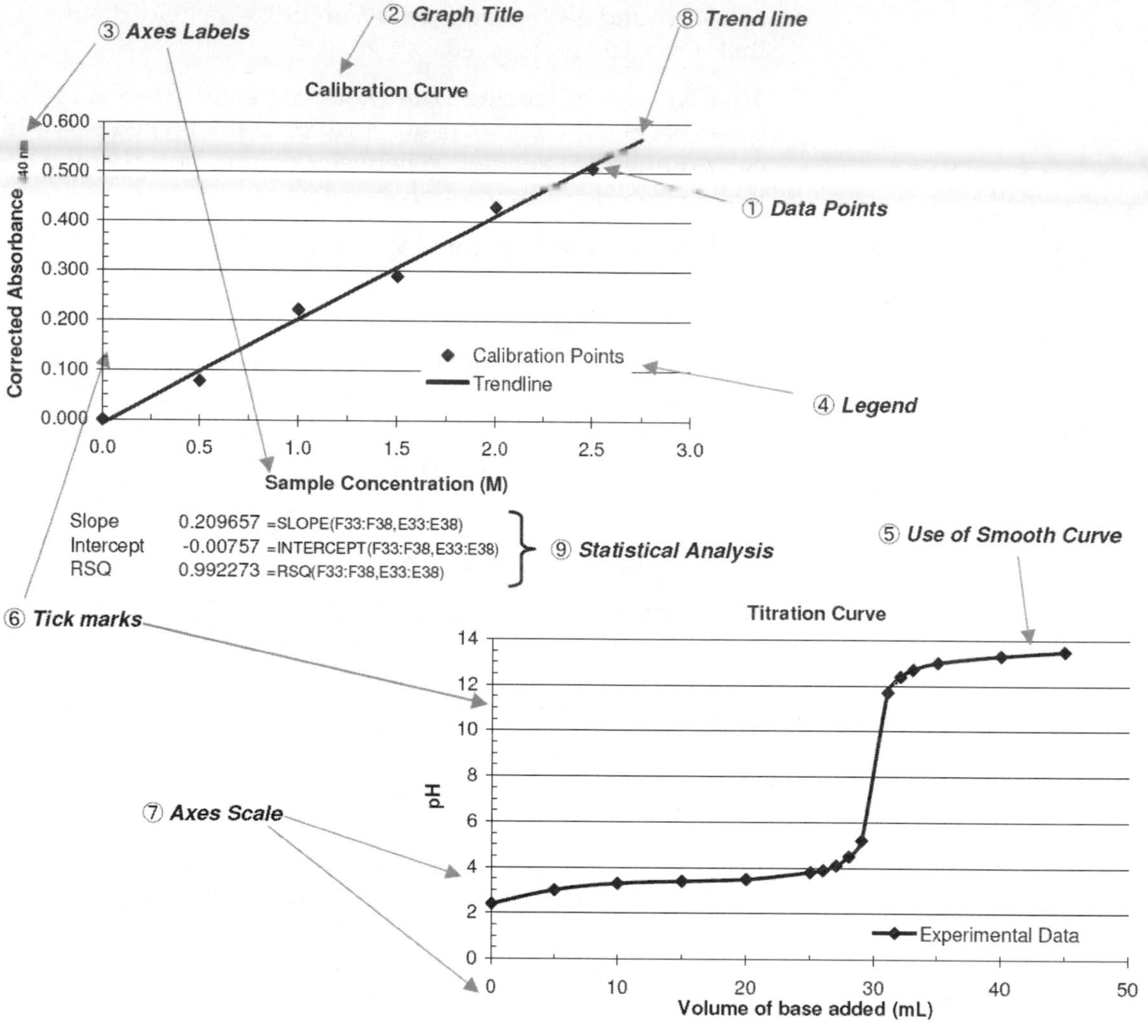

② Graph Title
③ Axes Labels
⑧ Trend line

Calibration Curve

Corrected Absorbance 540 nm
0.600
0.500
0.400
0.300
0.200
0.100
0.000

① Data Points

◆ Calibration Points
— Trendline

④ Legend

0.0 0.5 1.0 1.5 2.0 2.5 3.0
Sample Concentration (M)

Slope	0.209657	=SLOPE(F33:F38,E33:E38)
Intercept	-0.00757	=INTERCEPT(F33:F38,E33:E38)
RSQ	0.992273	=RSQ(F33:F38,E33:E38)

⑨ Statistical Analysis

⑤ Use of Smooth Curve

⑥ Tick marks

Titration Curve

pH
14
12
10
8
6
4
2
0

⑦ Axes Scale

◆ Experimental Data

0 10 20 30 40 50
Volume of base added (mL)

In Excel™, "Chart Wizard" will guide you through the procedures needed to construct a graph. After preparing a table of your experimental data on the spreadsheet, click "Chart Wizard Icon"* at the top in order to start "Chart Wizard." ***If you don't see the icon, click on "Add or Remove More Buttons" (down arrow) to select the icon to be displayed.**

① Your experimental data should usually be represented as "points" using markers, not a connected line. (There are some exceptions to this. See ⑤.) Thus, it is generally a good idea to start constructing your graph by choosing " XY scatter" with a dot sub-type at the first step in Chart Wizard (left figure).

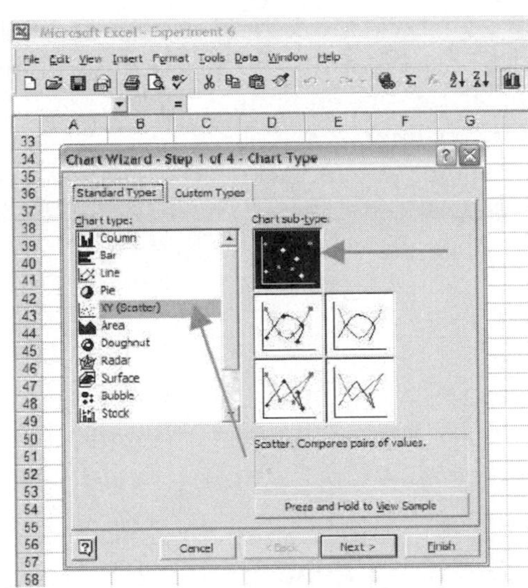

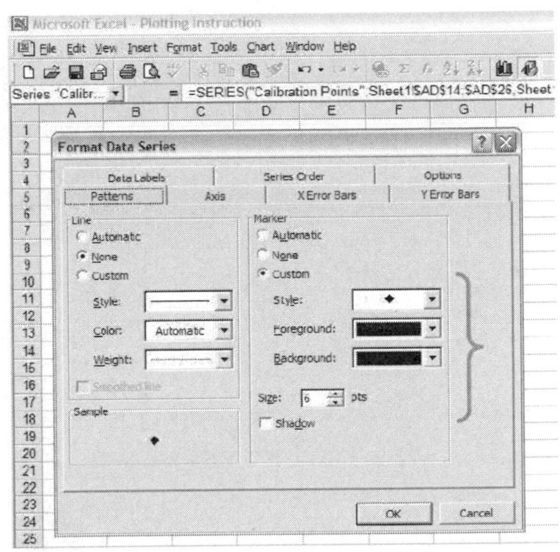

If desired, you can also change the style/size/color of the marker shown on your graph after finishing Chart Wizard. Bring up the "Format Data Series" window (right figure) by double clicking your actual graphed data point, and then select the "Patterns" tab to make the desired change.

At Step 2, you will be asked to select the range for your series. Select the range for each variable directly from the table of your data on the spreadsheet.

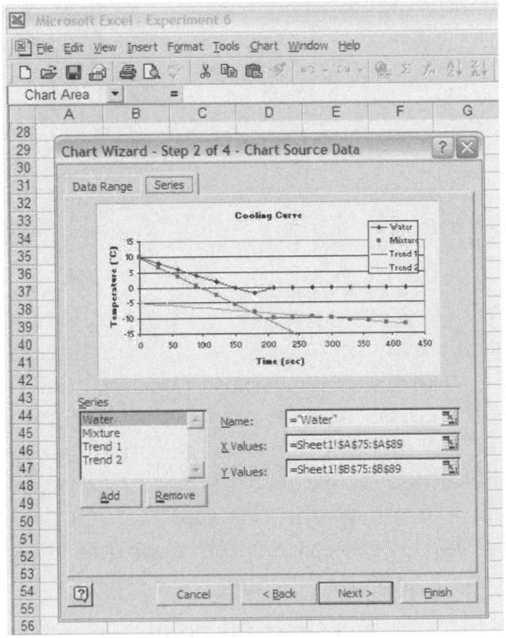

If you have "multiple series" as shown here, you do not have to complete this step all at once. You can always re-enter* to the Chart Wizard after the first completion of your graph.

***First, select your graph by clicking anywhere on your existing graph, and then clicking the "Chart Wizard Icon" to re-enter.**

② ③ ④ A Chart title, axes labels and legend are needed for each graph in your report. Step 3 in "Chart Wizard" lets you add these features to your graph. Again, you can always re-enter to Chart Wizard if you need to modify or add these features later.

In some cases (for instance, when you need to display temperature and time data), you will require specific units. These should be included in the axis labels.

- Temperature (°C)
- Time (sec)
- Time (min)

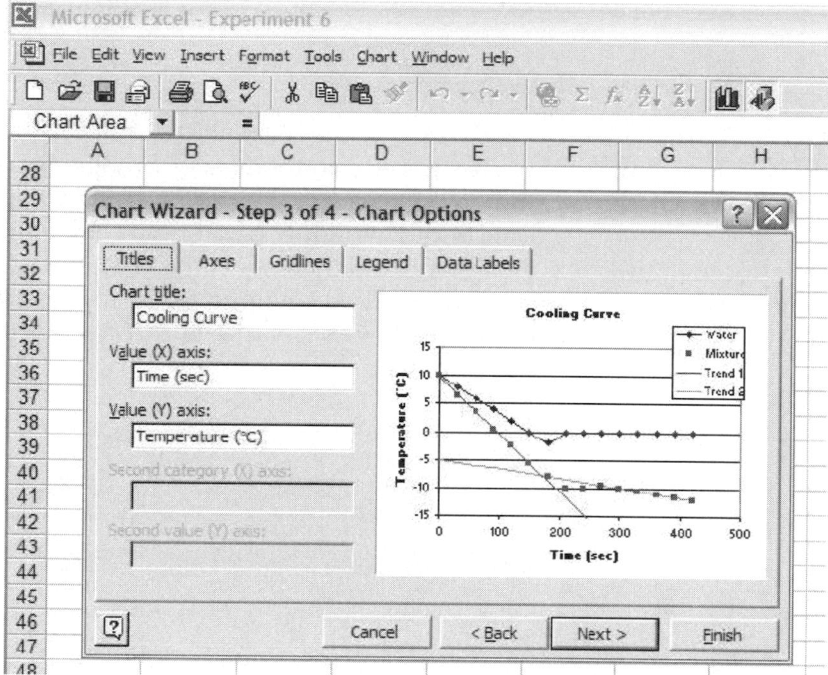

⑤ When your experimental points show a complex continuous curve, as in "pH Titration Curve," where no single polynomial function can provide a best fit of the data, you will need to use the "smooth curve" option.* The smooth curve can be drawn by either selecting "XY scatter" with a smoothed line sub-type as the first step with the Chart Wizard, or a line can be added (with the smoothed line box selected) to the existing markers on your graph by the "Format Data Series" window previously discussed in ①. ***Still, your experimental points should be presented with markers.**

⑥ Having "tick-marks" on your graph helps both you and your readers read off any experimental points on your graph for data analysis. When you construct your graph through Excel's Chart Wizard, "Major tick marks" are added automatically. However, for greater accuracy, you should consider adding finer "Minor tick marks" as well. Do this by double clicking any value on your *Y*-scale (or *X*-scale). The "Format Axis" window will appear and you will be able to make the necessary choices to provide the desired divisions on the axes.

Click on the "Patterns" tab (left figure), select the style for "Minor tick mark type." The default option, "scaling in axes," including amplitude of both major and minor tick marks, is adjusted automatically, but if desired, you can select the "Scale" tab (right figure) to adjust the scaling of your tick-marks (major or minor unit) for better presentation.

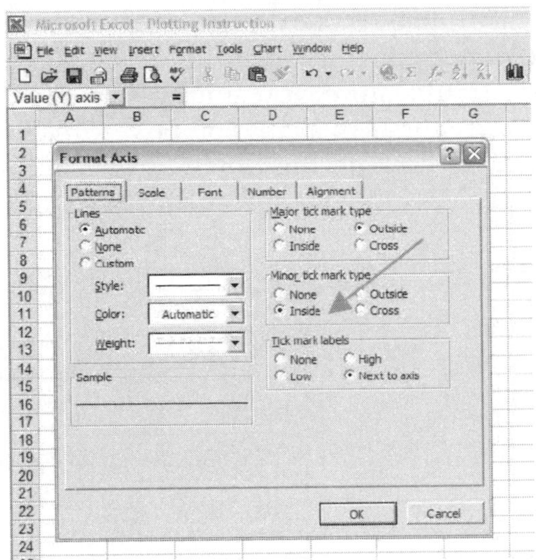

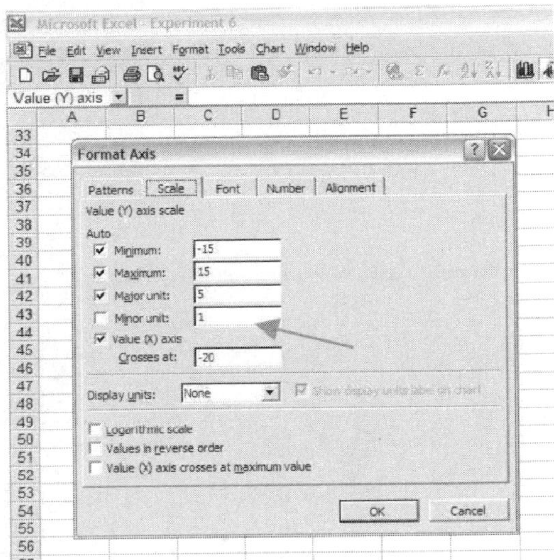

⑦ In the "Format Axis" window explained above, you can also adjust minimum / maximum of your axis under the "Scale" tab. Again, scaling is adjusted automatically by default, but can be adjusted manually* for better graphical presentation. Adjust the min/max of each axis so that your data points are optimally presented in a plotting area. ***Even if you need to adjust only the minimum or only the maximum, make sure to *uncheck* "Auto" boxes for both, otherwise auto-scaling will still take effect.**

⑧ When your experimental points show a linear trend, you should apply a trend-line (Fitted Linear Regression Line). In a simple case, as in "Calibration Curve," Excel's "Add Trendline" feature is useful. However, in a slightly more complicated case, such as when you wish to apply trend line to only a limited range of your data points, you need to perform linear regression manually.

i. First, compute "slope" and "intercept" for your data range (see [9] for detail).

ii. Create a column for your new Y to be calculated next to your original Y.

iii. Use slope and intercept, calculate new Y for each of your original X value.

iv. Add this new set of values (your original X and calculated Y) as a new series to your existing graph by re-entering to "Chart Wizard."

v. You can then change markers to a line from the "Format Data Series" window.

⑨ Mathematical functions available in a spreadsheet program will allow you to explore complex mathematical relationships with ease. For example, the "slope" computed with Excel's math function returns the value of the linear regression results through the given data range, which involves a far more complicated mathematical calculation than simply finding the slope based on two data points.

To employ this useful feature, you simply start by typing "equal sign (=)" as you normally do for any calculation on a spreadsheet. Then, select the desired function by clicking the down-arrow next to the "Function Box" at the upper left corner of your spreadsheet (see left figure).

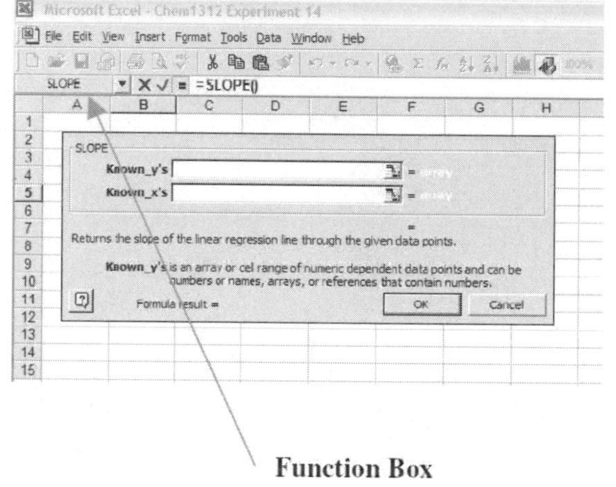

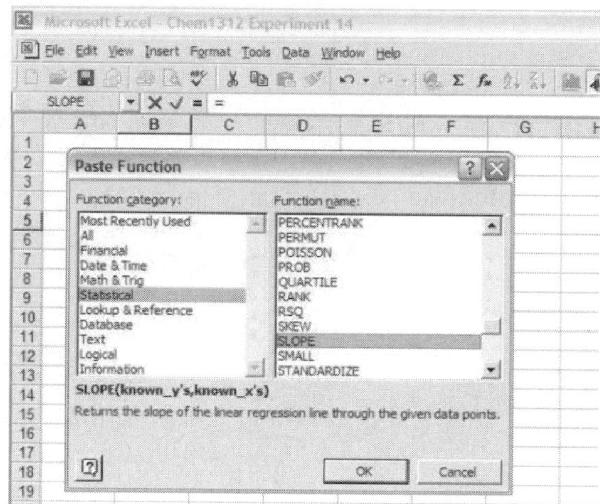

If you don't see a desired function, then select "More functions." The "Paste Function" window will appear (right figure), revealing more available choices.

While "slope" and "intercept" are probably the most frequently used math functions for a graphing purpose, many useful math functions for statistical analysis are also available. For example, "RSQ" (square of Pearson product moment correlation coefficient*) helps you analyze how strongly your experimental points exhibit a linear association. Such statistical value can be also used to support the validity of your experimental data. ***By definition, the closer the RSQ value is to 1, the stronger the linear association.**

Any value computed via the program's math function must be accompanied by a pasted formula to show what function was used and the range of data used. Simply copy and paste the formula from the "formula bar" and place it next to your computed value.

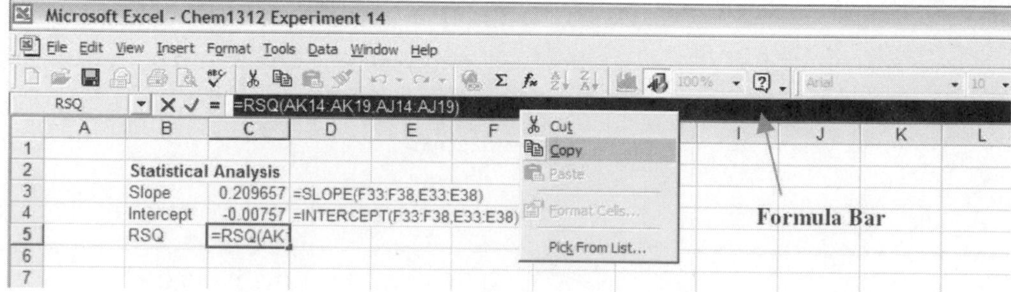

Note: You can convert your formula to "text" by placing an apostrophe (') in front of the equal sign after pasting.

Since these pasted formulas show "cell coordinates", you need to setup your sheet so that "Row and Column headings" will appear on your printed spreadsheet. Bring up the "Page Setup" window (left figure) by selecting "Page Setup" under "File" menu. Click on "Sheet tab", and then check both "Gridlines" and "Row and column headings".

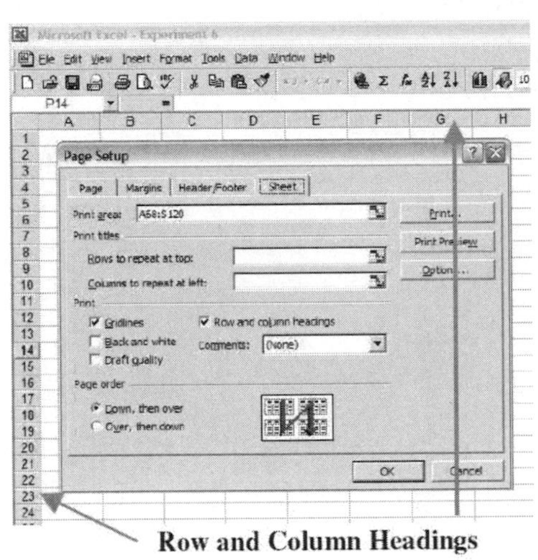

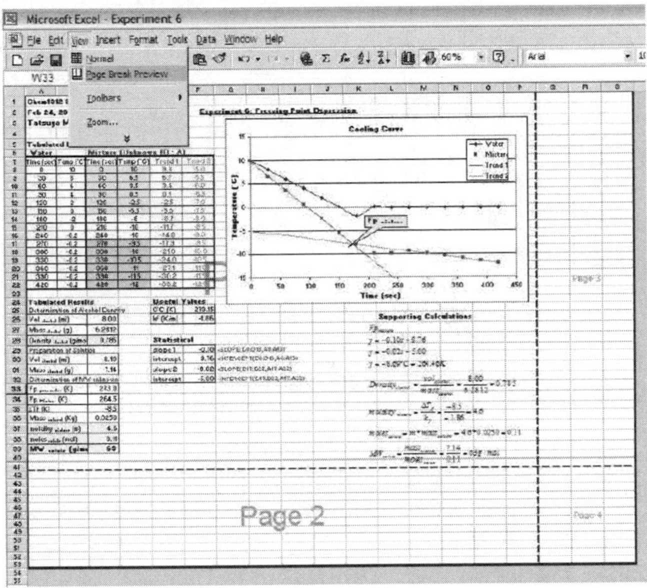

Switching view to the "Page Break Preview" (right figure) will give you an instant display of overall layout, and also let you adjust the printing range. Adjust borders so that all of the contents in your spreadsheet will be presented optimally when printed.

Further details on many useful features, including those discussed above, can be found from "Help" in Excel™. Here, you will find descriptions and "how-to" for features in Excel™, and also explanations on how each math function computes values.

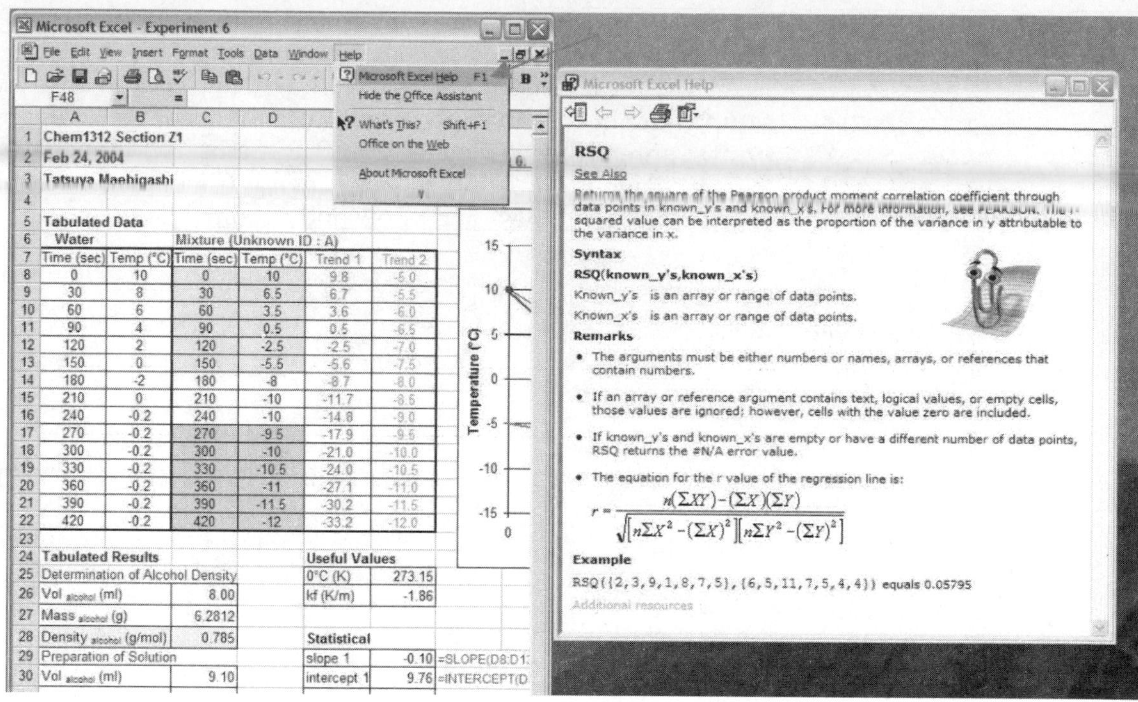
Microsoft Excel - Experiment 6
File Edit View Insert Format Tools Data Window Help
Microsoft Excel Help F1
Hide the Office Assistant
What's This? Shift+F1
Office on the Web
About Microsoft Excel
Chem1312 Section Z1
Feb 24, 2004
Tatsuya Maehigashi
Tabulated Data
Water
Mixture (Unknown ID : A)
Time (sec) Temp (°C) Time (sec) Temp (°C) Trend 1 Trend 2
Tabulated Results
Determination of Alcohol Density
Useful Values
0°C (K) 273.15
kf (K/m) -1.86
Statistical
slope 1 -0.10 =SLOPE(D8:D1
intercept 1 9.76 =INTERCEPT(D
Preparation of Solution
Temperature (°C)
Microsoft Excel Help
RSQ
See Also
Returns the square of the Pearson product moment correlation coefficient through data points in known_y's and known_x's. For more information, see PEARSON. The r-squared value can be interpreted as the proportion of the variance in y attributable to the variance in x.
Syntax
RSQ(known_y's,known_x's)
Known_y's is an array or range of data points.
Known_x's is an array or range of data points.
Remarks
The arguments must be either numbers or names, arrays, or references that contain numbers.
If an array or reference argument contains text, logical values, or empty cells, those values are ignored; however, cells with the value zero are included.
If known_y's and known_x's are empty or have a different number of data points, RSQ returns the #N/A error value.
The equation for the r value of the regression line is:
Example
RSQ({2,3,9,1,8,7,5},{6,5,11,7,5,4,4}) equals 0.05795
Additional resources

APP B

Properties of Water

Table B.1 Density of Water at Various Temperatures

T(°C)	d(g/mL)	T(°C)	d(g/mL)
0	0.99984	26	0.99679
5	0.99997	27	0.99652
10	0.99970	28	0.99624
15	0.99910	29	0.99595
16	0.99895	30	0.99565
17	0.99878	31	0.99534
18	0.99860	32	0.99503
19	0.99840	33	0.99471
20	0.99821	34	0.99437
21	0.99800	35	0.99403
22	0.99777	40	0.99222
23	0.99754	45	0.99022
24	0.99730	50	0.98804
25	0.99705	55	0.98570

Table B.1 Vapor Pressure of Water at Various Temperatures

T(°C)	Vapor Pressure (torr)	T(°C)	Vapor Pressure (torr)
0	4.6	31	33.7
5	6.5	32	35.7
10	9.2	33	37.7
15	12.8	34	39.9
20	17.5	35	42.2
21	18.6	40	55.3
22	19.8	45	71.9
23	21.1	50	92.5
24	22.4	55	118.0
25	23.8	60	149.4
26	25.2	65	187.5
27	26.7	70	233.7
28	28.3	80	355.1
29	30.0	90	525.8
30	31.8	100	760.0